Felix Publishing 2018
www.felixpublishing.com.au
email: info@felixpublishing.com
Print copies available from publisher.

A Dangerous Planet
Part of the Series **Adventures in Earth Science**
Other books in the series include:
>   Exploration Science (Field Geology & Mapping)
>   Riches from the Earth (Minerals, Mining & Energy)
>   Changing the Surface (Erosion & Landscapes)
>   Rocks – Building the Earth
>   Fossils – Life in the Rocks
>   Through Sea and Sky (Oceanography & Meteorology)
>   Beyond Planet Earth (Astronomy)

2018 digital book release
ISBN: 978-0-9946433-5-3
Print Edition
ISBN: 978-0-994643377
Author: Dr Peter.T.Scott

All illustrations, photographs and videos by the author unless stated. Cover photo: A lava window through a thin crust on Mauna Ulu volcano Hawaii, 1970 (Photo: USGS). Design based on that of AJS Creative, Brisbane.

Registration:
Thorpe-Bowker +61 3 8517 8342
email: bowkerlink@thorpe.com.au

FELIX
PUBLISHING

To my Grandchildren who are
yet to find their own adventures.

A

DANGEROUS PLANET

Dr. Peter T. Scott

First released 2017

FELIX
PUBLISHING

About The Author

Dr. Peter Scott is an award-winning teacher of Earth Science of over forty years' experience in both Secondary and Tertiary Education. He holds a Bachelor's Degree, two Master's Degrees and a Doctorate including many years on his own research in locating and correlating coal measures. He has travelled extensively throughout seven continents and has visited many places of interest including Antarctica, the Andes, the Amazon, North Africa, volcanic islands of the Pacific and Asia, the USA, northern Europe and remote Australia.

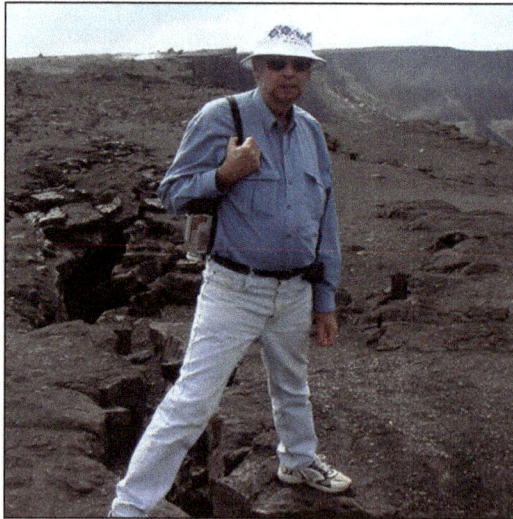

Dr. Scott, in the larger crater of Kilauea volcano, Hawai'I, 2011

Table of Contents

Chapter 1: The Surface in Motion

1.1 Introduction

The dynamic nature of the Earth has been known for a long time. Volcanoes and earthquakes have been part of the life of many people in many countries around the globe. In Europe, where the new sciences were developing from the 17th century, scientists observed that the basalt rock structures in Ireland, Scotland and elsewhere were identical in composition and structure to those fresh lava flows observed in active volcanic areas of Italy and Iceland.

As people became more aware of the rocks around them, they also noticed that some parts of the land surface had risen slightly out of the sea to give flat rock platforms and others had sunk below it inundating farmland and villages. Moreover, layers of rock, or strata, which were normally horizontal, had somehow become tilted, folded or even cracked and displaced. The Earth had shown signs that it had moved, very slowly in places, to cause these structures, but at other times, during earthquakes and land collapse, very rapidly. Volcanic activity also demonstrated that the Earth was still in the process of construction. In time, geologists became aware that the Earth's surface was in continual motion, mostly slow and very imperceptible, but at times fast and destructive. From a human point of view, these movements and the volcanoes and earthquakes which they produced, showed that planet Earth was a fragile and dangerous place.

Figure 1.1: Looking across the University of Edinburgh, Scotland towards Arthur's Seat, the old lava flow which was one of the first structures to be studied during the early foundation of geology at the university.

1.2 Moving Slowly

With the development of geology, it had been noted that the Earth's surface has been pushed up either as broad vertical movements called **epeirogeny**, from the Greek epeiros for land and genesis for creation, or as mountain-building called **orogeny**, from the Greek *oros* for mountain and genesis for creation, which is uplift on a very large scale caused by multi-directional forces.

In early thinking, the building of mountains and other earth movements were thought of as a simple maintenance of a surface equilibrium which balanced vertical movement upwards against the wearing down of mountain ranges by erosion. This general maintenance of a vertical equilibrium is called **isostasy,** from Greek *ísos* for equal and *stásis* for standstill. By the 20th century, evidence from new studies using the new theory of Plate Tectonics, showed that the whole surface of the planet consisted of separate slabs or plates which were in constant motion and that at different places, the interaction of plates caused uplift of mountain ranges, the sinking of oceanic landforms as well as volcanoes and earthquakes. Evidence for large-scale movement of the Earth's crust can be seen by:

- Fossils have been found at great altitudes suggesting that the environments in which these organisms once lived had been uplifted by forces within the Earth. Fossilized marine molluscs, such as the ammonites were to be found high in the Himalayan range, and dinosaur footprints have been found in the Andes Mountains.

Figure 1.2: Fossil Mollusc similar to those found in the Himalayas

Figure 1.3: Dinosaur footprints (?) and sedimentary ripple marks found in the high Andes of Chile near the Morado Glacier.

4

- Marine rock platforms, which have been originally eroded as flat surfaces below the low tide mark by wave action but are now permanently above sea level.

Figure 1.4: Sandstone cliffs and the eroded marine rock platform below. Caloundra, Queensland, Australia.

- Exposures of deeply-formed igneous rocks on the surface are caused by broad uplift accompanied by rapid erosion. This exposes rock which initially had been formed many kilometres below the surface from the cooling of molten rock within large igneous structures such as batholiths. Rocks such as granite with large crystals can now be found as groups of rounded boulders called **tors**.

Figure 1.5: Granite tors are the surface feature indicating a huge batholith of intrusive igneous rock which has been uplifted from several kilometres below

- Continued mountain-building in many of the world's great mountain chains such as the Himalayas, Rocky Mountains and the Andes can now be measured directly in real time. This is done by accurate surveying techniques on land and by earth satellites using triangulation in geodetic networks. In this method, several positions on the Earth's surface are connected as points in a triangular pattern, so that very accurate data giving their altitude can be compared. Accurate land surveying methods have been used since the 19th century, when **Sir George Everest** (Welsh: 1790-1866), measured the height of the Himalayas in northern India. More recently, satellites have used dual-band radar altimeters, augmented with GPS (Global Positioning System) data, to measure height by the reflection of two different radar waves off the terrain below.

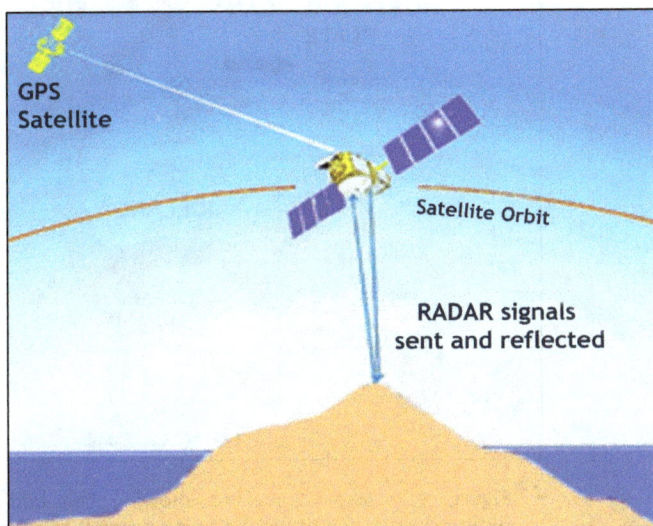

Figure 1.6: Diagram showing how heights of mountains can be measured using satellites

- Direct measurement can be used across mid-ocean ridges such as in the Thingvellir National Park, in southwestern Iceland, where the land surface is spreading apart by up to two centimetres per year. This movement can be measured using **LASER geodometers**, which accurately measure the time of reflection of a LASER beam off a target reflector positioned across the rift valley from the source of the beam. This system is also coupled to a system of continuous GPS networks so that both geodimeter and target reflector position can be very accurately known;

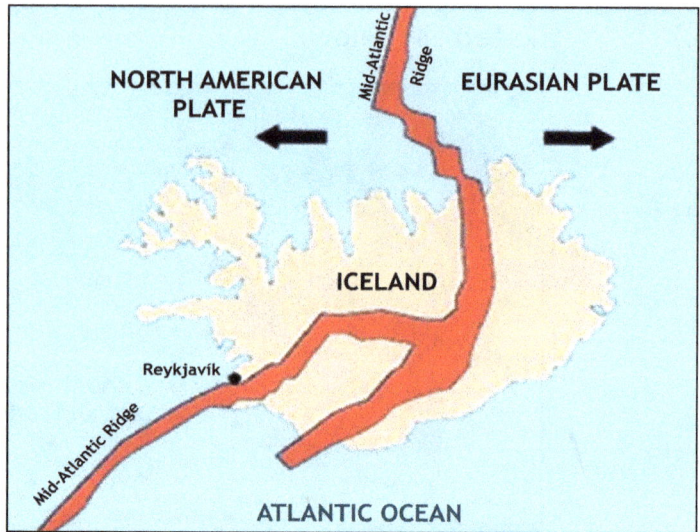

Figure 1.7: Map of Iceland showing the mid-ocean ridge extending through the island and causing it to split apart

1.3 The Nature of Rock

Before one can understand how the rocks of the Earth's crust behave when subjected to force and movement, one must understand the physical nature of rock material. In addition, rocks will behave differently in response to how the forces are applied. A force which is applied suddenly will cause rock to act like a brittle material and so it will suddenly break. If this same force is applied over a very long period of time, the rocks will act in a more flexible manner and will bend and deform but remain intact.

Different rocks have different strengths because of their component minerals, how these are interlocked together, the type of bonding or cement which holds them together and any fluid content. They behave differently under equal conditions of stress, that is, applied force per unit area, such that weaker rocks can be easily be deformed or fractured and stronger rocks less so. Strain, or the amount of deformation a rock undergoes, is measured as the ratio of the increase in length, surface area or volume of the rock, to that of the original dimensions of the rock body.

Stress (σ) = force applied / area
(measured in newtons/square metre)

Stress can be caused by forces being applied in several ways:

- tensional stress which stretches rock

- compressional stress when rock is pushed inwards from all sides

- shear stress when the force is applied off-centre and there is slippage and translation within the rock

- confining stress when forces are applied equally from all directions.

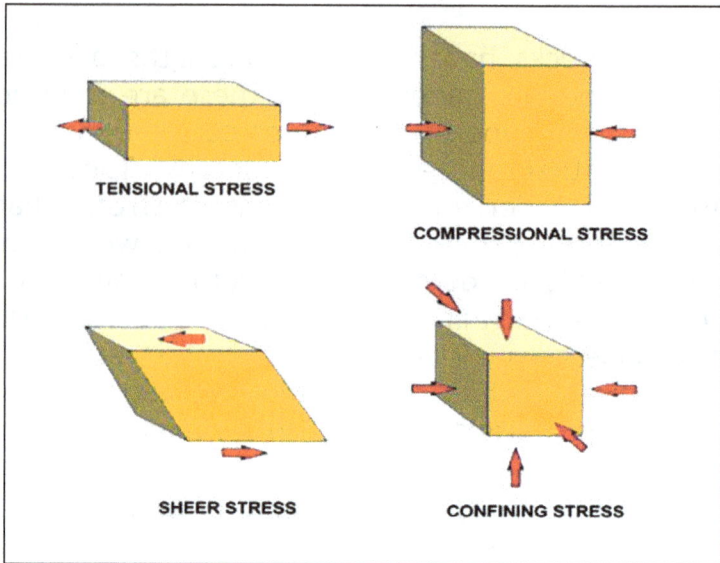

Figure 1.8 Different types of stress in rock

Strain is measured as a ratio of the increase in deformation divided by the original dimensions of the rock length, area or volume:

Strain (ε) = <u>increase in dimensional deformation (δD)</u>
original dimension (D)

The strength of the rock may also vary if conditions change. Strength is generally <u>reduced</u> by:

- raising the temperature

- applying large forces for short periods

- applying smaller forces for long periods

- increasing the amount of fluids in the rock

Rocks tend to behave in two major ways when stresses are applied. They may behave in a **ductile** manner and appear to flow and fold or they may behave in a **brittle** manner and suddenly break. When rock suddenly breaks, large cracks or joints may form and if there is movement along these joints they become faults.

Graphically, the nature of the type of rock structure produced by different conditions of stress and strain can be shown as:

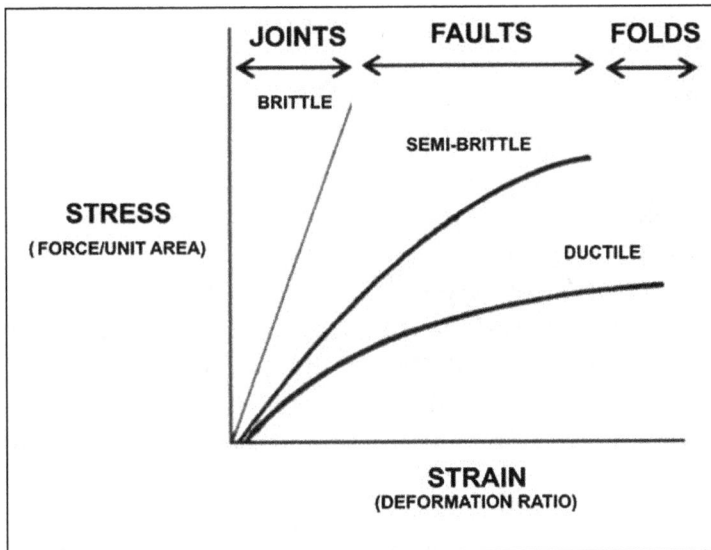

Figure 1.9: Graph showing the stress and strain upon rock and the structures which result.

From this graph, it can be seen that brittle rocks do not have much deformation and so quickly break giving joints and faults, whilst ductile rocks begin to deform easily and fold, reaching their **elastic limit,** where the curve starts to bend, soon after stress is applied

Within the elastic limit, or the range in applied force in which a deformed rock will return to its usual shape after the force has been removed:

Stress is proportional to Strain

This relationship is known as **Hooke's Law.**

1.4 Brittle Structures - Joints & Faults

Brittle structures are joints, or fractures, and faults which occur in rocks in response to high stress. This applied stress overcomes the rock's strength so that it breaks along planes of weakness. Usually very little strain is involved, so there is little deformation along or around the break. The stress producing the brittle structure could be caused by such natural forces as burial pressure of the weight of rocks above, uplift of igneous bodies, and movements of continental plates transmitted throughout the rock of the continent or sea floor.

- **Joints** are planar, that is flat, 3-dimensional, fractures in rock along which no movement has occurred. In size, these vary from the microscopic defects seen within individual minerals, to the major fractures in the Earth's crust.

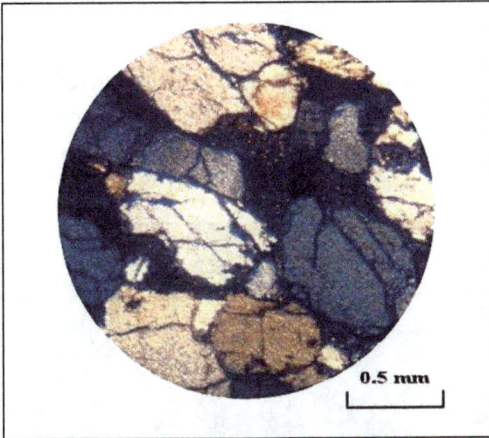

Figure 1.10: Small fractures in quartz grains seen in thin-section of sandstone

Figure 1.11: Large fractures in the rock wall of the valley the Franz Josef Glacier, New Zealand -note the size of the people at its base

13

Some of the main types of joints or **fractures** include:

- Axial fractures which are parallel joints due to tensional forces pulling the rock apart. These are often caused by the rapid unloading of rock above by erosion.

Figure 1.12: Small axial fractures in a fine sandstone, Kholo, Queensland, Australia.

- Sheeting joints are axial fractures produced at or near the surface by the reaction forces within rock as the weight of the rocks above is removed by erosion and the solid rock below rebounds. This can be seen in sheeting similar to exfoliation, as the peeling of layers off the surfaces of hard rocks such as granite

Figure 1.13: Sheeting joints in granite near Mt. Lyell, Tasmania, Australia

- Orthogonal joints occur when the cracks within the rock occur at mutually perpendicular angles to each other such as is seen in the rock limestone.

Figure 1.14: Orthogonal joints in limestone of the Barran, County Clare, Ireland which have been enlarged by weathering

- Conjugate Joints consist of pairs of joints inclined at an angle of about 60 degrees. The compressional forces which fractured the rock are shown in the diagram below. An analysis of these joints can indicate the direction from which the compressional forces have come.

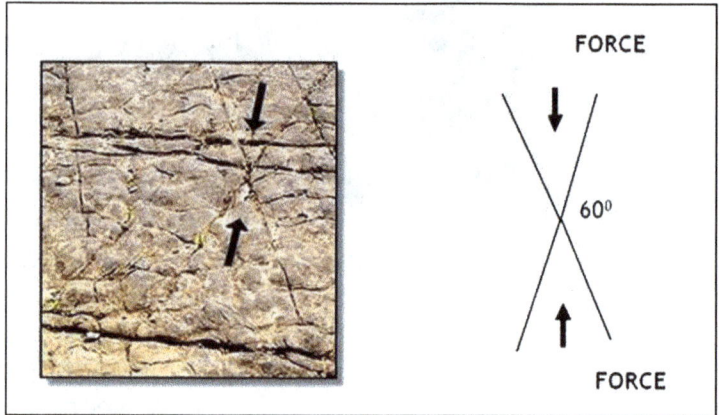

Figure 1.15: Conjugate joints (left) and the relationship to the force producing them

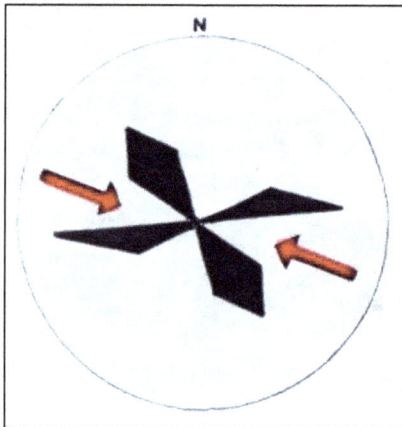

Figure 1.16: A Rose diagram showing the results of the analysis of over 1000 joints within a bed of rock. The geographical direction of each joint is found and graphed such that their number is represented by the distance out from the centre and their bearing by the 360^0 compass rose. Here, the compressional forces causing the joints came from the west north west and east south east.

- Columnar joints are formed from contraction around cooling centres when molten rock cools. This produces polygonal and usually hexagonal columns in volcanic rocks such as basalt. Slabs cut horizontally through these hexagons were used by the Romans to pave their roads.

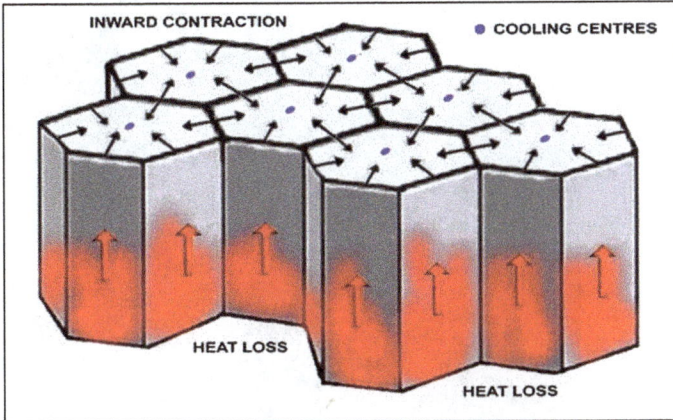

Figure 1.17: A diagram showing how hexagonal columns are formed in lava as it cools and contracts inwards

Figure 1.18: Hexagonal columns of basalt – Giant's Causeway, Northern Ireland.

- **Faults** are structures found in semi-brittle rocks in which there has been movement along the plane of the fracture. The displacement could be a few millimetres or many thousands of kilometres in length.

Figure 1.19: The San Andreas Fault which runs for 1300 km. through the west coast of the USA (Photo: Robert E. Wallace, USGS)

Faults are often the result of forces associated with the movement of the Earth's crust. They are associated with earthquakes, often accompany violent volcanic activity or they may simply be a sudden fracture of the rock due to the forces transmitted through the crust.

The main descriptive features of a fault are the:

- Fault block which is the section of rock on either side of the fault.

- Fault plane which is the two-dimensional face of the fault. Below ground, there may be a fault breccia of crushed rock material formed by the friction of the moving blocks.

- Fault scarp which is the escarpment or cliff which may result from the vertical displacement of the fault. This may not be obvious in older faults where it may have been eroded or covered with a scree slope. Some faults move sideways without much vertical movement so this feature will be small or absent. Where the fault scarp is bare rock, **slickensides**, or eroded grooves may be seen.

Figure 1.20: Slickensides seen in serpentine mineral on a rock surface near the Morado Glacier in the Andes of Chile.

- Fault dip or fault angle of dip which is the angle that the fault plane makes to the horizontal.

- Fault strike which is the geographical bearing or compass bearing of the line of the fault on the surface.

- **Foot wall** which is the section or block that is below the fault. It is the solid footing which does not move.

- **Hanging wall** which is the section or block above the fault and is the part which moves.

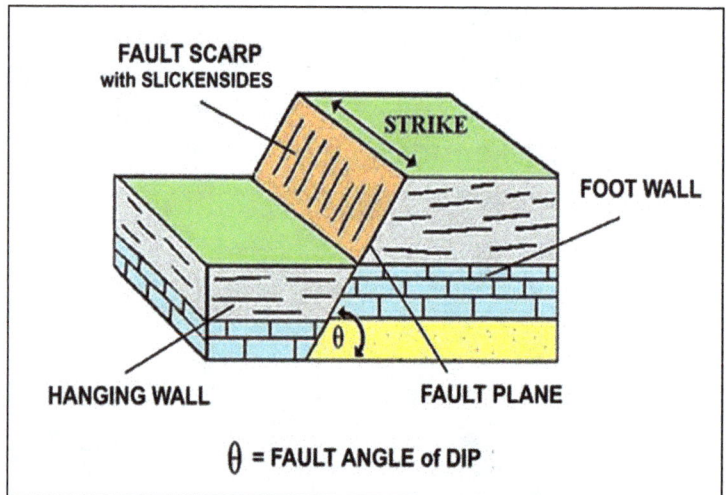

Figure 1.21: Diagram showing the main parts of a typical fault

There are two other dimensions which are often given when describing the characteristics of a fault in respect to how the fault has moved (its slip). These are heave and throw:

- **Heave** is the horizontal displacement of any **marker bed** or any linear feature seen on both sides of the fault in vertical section. In this diagram it is the blue limestone – or more accurately, its top surface shown by the blue line.

- **Throw** is the vertical displacement of the marker bed.

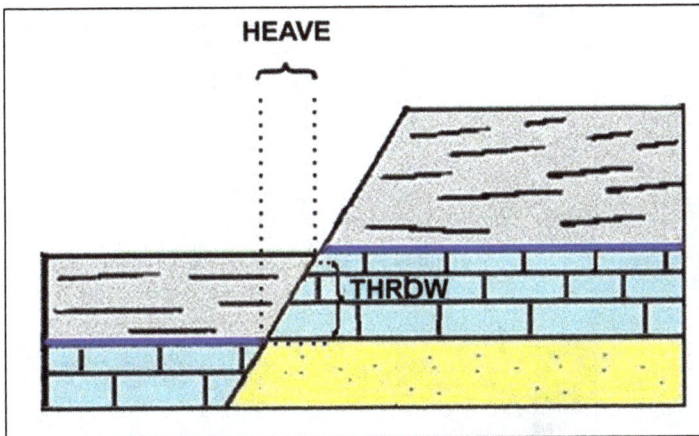

Figure 1.22: Heave and throw of a fault

Depending upon their direction, the forces which produce the fault may be tensional, which is pulling the rocks apart or compressional which are pushing inwards. The type of applied force will determine the type of fault which will result and several major types of faults can be recognized. These include:

- Normal or tensional faults are formed as blocks are pulled apart and the hanging wall slips down the dip or downward direction of the fault. Because of this, these are often called dip-slip faults.

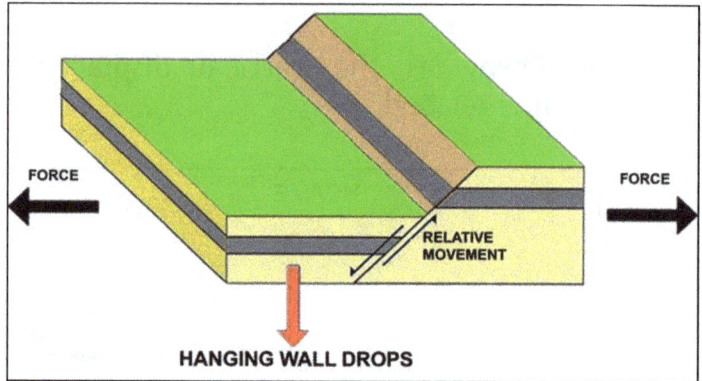

Figure 1.23: Diagram of a normal fault

Figure 1.24: Normal fault shown in sandstones and shales in a road cutting

- Reverse faults are formed as blocks are pushed together and the hanging wall is pushed up over the footwall. They are also called compressional or thrust faults. The apparent sharp edge of the fault scarp is quickly eroded away and so it is not so severe and may look like a simple hill slope. However, in an extended view, especially with aerial or satellite imagery, the line of the fault can often be clearly seen.

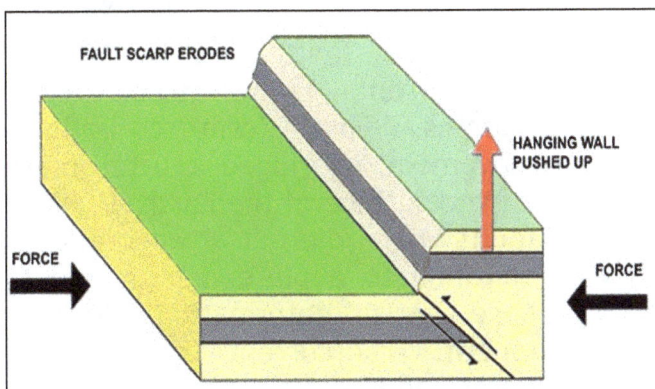

Figure 1.25: Diagram of a reverse fault

- Transcurrent faults occur when the blocks have been displaced sideways with little or no vertical movement. As the movement is along the direction of the fault line, or the strike of the fault, these are also called strike slip faults.

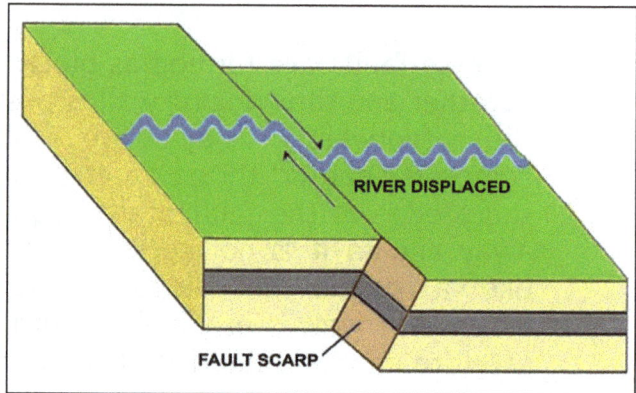

Figure 1.26: Diagram of a transcurrent fault

- Transform faults are a special type of transcurrent fault associated with plate margins. They form to allow horizontal movement of ridged crustal plates over the curved surface of the Earth. Unlike transcurrent faults which die out at each edge, these faults continue to a plate margin and they tend to have equal deformation along their entire length. Transcurrent faults are usually more deformed in their centres.

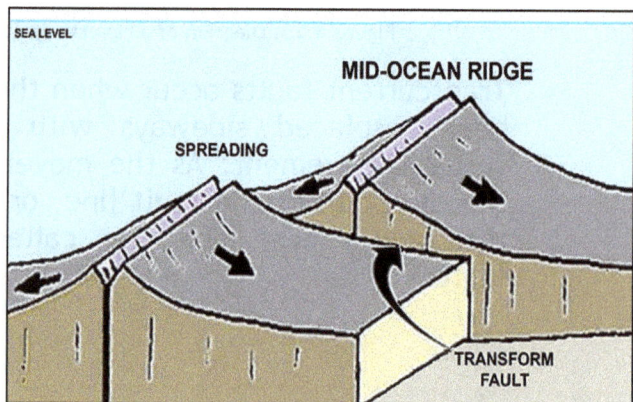

Figure 1.27: Diagram of a transform fault

- Hinge faults are formed as a fracture opens along the line of the fault more on one end of the fault than the other, much like a door being opened, such that the throw or separation of the beds, increases along the fault line.

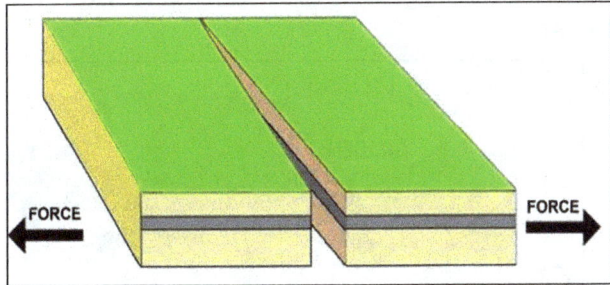

Figure 1.28: Diagram of a hinge fault

- Oblique-slip faults are a combination of horizontal movement along the fault line and vertical movement downwards. Many faults have some components of both vertical and lateral movement.

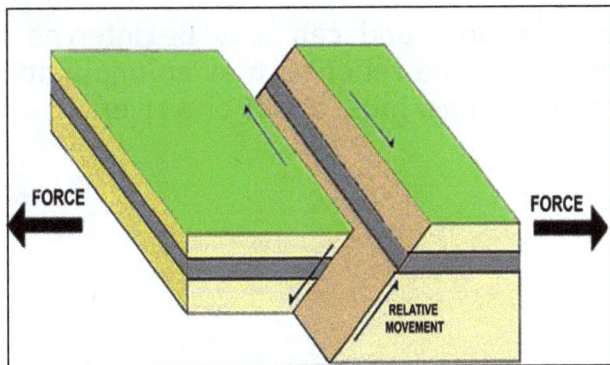

Figure 1.29: Diagram of an Oblique-slip Fault

In very badly fractured regions, there may be several fault types together over large distances. Many major faults are often complexes of several parallel fault lines, often with adjacent faults at right-angle to them. Combinations of faults also produce large, faulted valleys called rift valleys or **grabens** from the German for grave and uplifted block mountains or **horsts.**

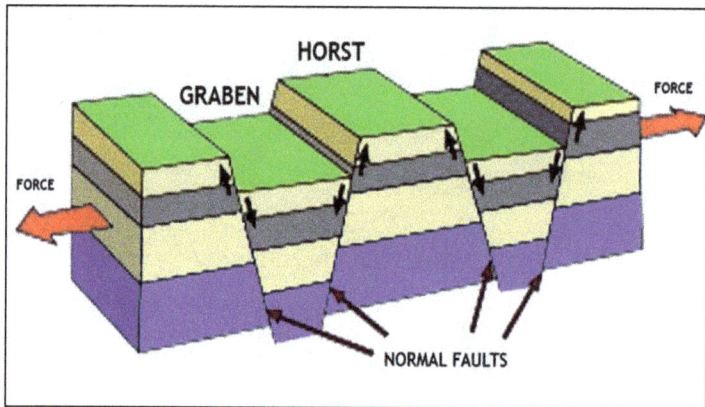

Figure 1.30 A horst and graben system

In the field, faults are often obscured by weathering and erosion, and can only be inferred by a sudden change in the relief such as an unusually straight line of mountains, hills, a cliff or a river.

Figure 1.31: Sometimes the fault line is dramatic – flat plains and the sudden rise of the New Zealand Alps along the Transalpine Fault. (South Island, New Zealand)

1.5 Landslides

When rock is fractured or soil is saturated with water and the **slope gradient** is steep, it may move downhill. This could be slow and produce a rippled surface known as **soil creep** or it may be sudden and fast as with landslides. Soil creep, especially in places of clayey soil is seen by the tilting of fences, telegraph poles, buildings and parallel fractures in roads. Any movement of soil or rock, generally termed **regolith**, downhill is called **mass-wasting**.

Mass-wasting may be triggered by any vibration such as construction vibrations and the use of explosives, volcanic eruptions nearby, earthquakes and sudden injection of large amounts of water. Collapse of layered rock (strata) is also helped when the rock layers are tilted downslope, if the beds are separated by porous clays which can absorb water and if considerable load from the construction of roads or buildings is placed upon the surface layers. In such a case, the layers will come apart and form a **rock slide** as sections move downhill, sliding over each other in a **translational** motion.

Slumping also will occur on steep slopes when a load is applied and/or the rock and soil is saturated with water. This mass-wasting involves a rotational movement as the regolith parts and slips downhill in an outward curve or concave collapse.

Rock falls are simpler in their motion. Here, the slope may become too steep to hold any fractured or loose rock so it simply drops downslope. This is especially common on the banks of rivers and sea cliffs where undercutting will allow the sudden fall of rock above.

Debris flows are especially destructive when large amounts of regolith on steep slopes are saturated with water from rain or melted snow. **Lahars** are mud-water flows which come from the steep slopes of erupting ash volcanoes are particularly dangerous and potentially fatal.

Sudden mass-wasting can also occur underwater when parts of the coastal Continental Shelf or Slope suddenly gives way and slides downslope as a **turbidity current.**

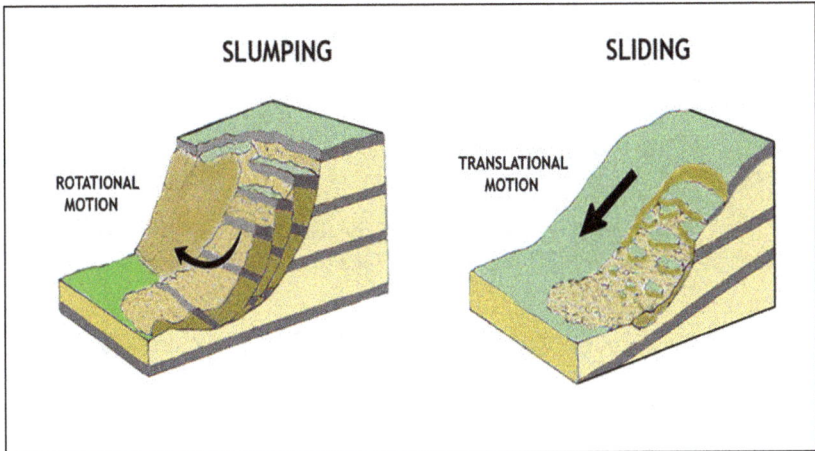

Figure 1.32: Slumping and Sliding as examples of mass-wasting

1.6 Ductile Structures - Folds

Folds are ductile rock structures, that is, structures caused by bending rather than breaking of rock. They are formed if the rock is able to flow when compressional stress is slowly applied to them. They may still be in a semi- solidified state, such as sedimentary rock which still contains considerable moisture and has not yet fully solidified, or they may be of a suitably high temperature so that they can be made to flow, when forces are applied slowly over a long period of time.

The capacity of a rock to behave like a fluid and flow is termed its **rheidity** and is measured as the time taken for deformation to exceed 1000 elastic limit deformations. If forces are applied for times greater than the rheidity of the particular rock, it will be deformed like a fluid (i.e. flow). Some common rheidities are:

- glacier ice - about two weeks

- salt (in underground domes) - about one year

- crustal rock- about 10^5 to 10^9 years

Often the folding also produces many small fractures as the rock strata are bent and the shape of each bed is distorted. Folds can be described in terms of the angles of their curved sides, or limbs and the orientation of their fold axis.

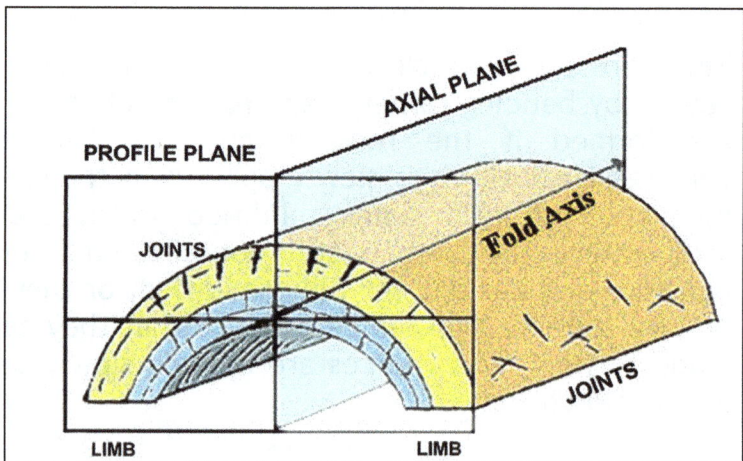

Figure 1.33: Diagram showing the main features of a fold

In terms of their profiles, Folds can be symmetrical, having limbs with the same angle of slope on the opposite sides of the fold axis, or asymmetrical with different angles on the opposite sides of the central axis.

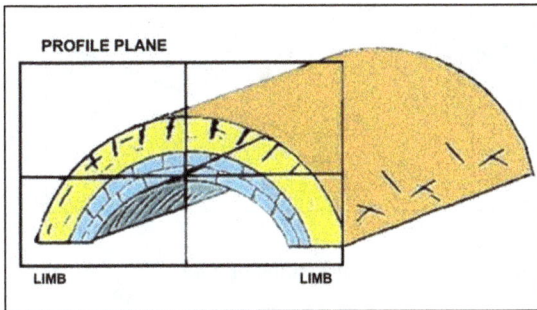

Figure 1.34: A symmetrical fold

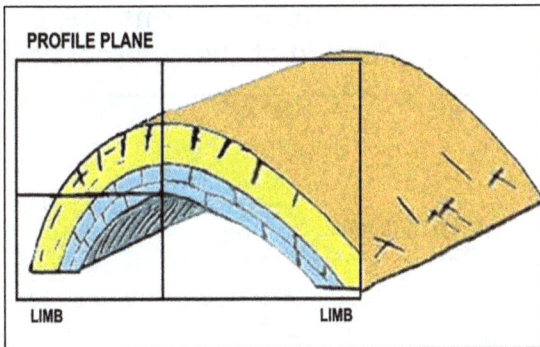

Figure 1.35: An asymmetrical fold

After the beds are folded, their fold axis may also suffer tilting if another force is applied in this new direction. This may produce a **plunging fold**:

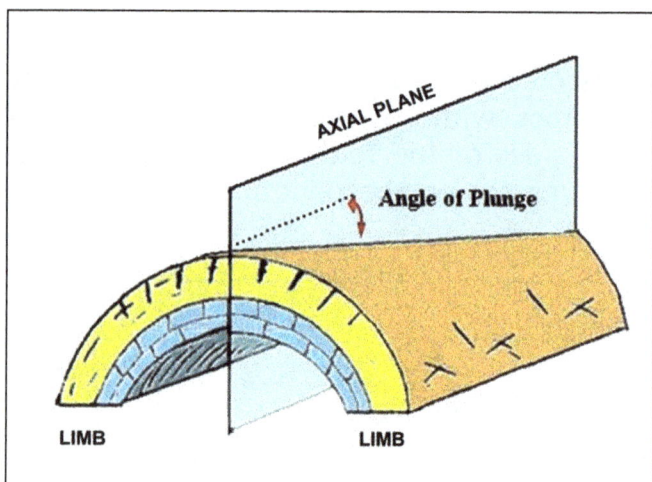

Figure 1.36: Diagram of a plunging fold

Folding comes in many different forms depending upon the severity and direction of the compression:

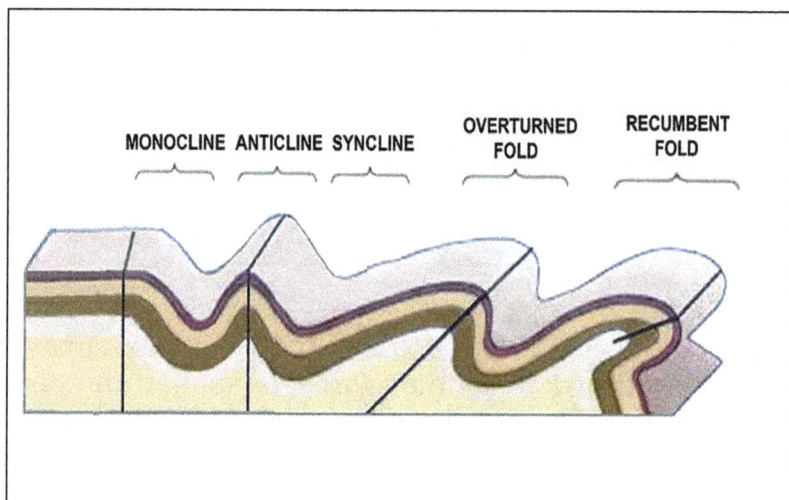

Figure 1.37: Diagram of the main types of folds

On a larger scale, a series of large folds may form rows of rounded hills or fold mountains. Under extreme compression, such as when one plate collides with or is subducted below another, there may be severe recumbent folding such that strata is completely overturned with the youngest layers now being <u>below</u> the oldest layers. If there is considerable erosion so that only a small part of the recumbent fold is exposed, it may show strata which appears to disobey the **Law of Superposition**. Thrust faulting is also associated with such extreme folding and wedges of partly folded rock may be pushed up as a **nappe**, French for tablecloth when it is pushed and wrinkled along a table top. There are several exposures of nappes in the eastern European Alps, the Carpathian Mountains further east and in the Balkan mountain range. Some of the terms associated with folding are:

- **Syncline** is an upward fold.

- **Anticline** is a downward fold.

- **Monocline** is a fold with one limb, usually at a steep angle. Instead of a second limb it may continue on to form a bench or a geological step or taper off as a long slope.

- **Overfold** which is also known as an overturned fold, is a fold with one limb that has a down dip of 90^0 or more. An extreme example of this type of fold occurs as a **recumbent fold** in which one limb folds beneath itself forming a complex fold system.

Fold systems can often be very complex and may involve faulting as well, depending upon the speed at which the folding has occurred. Usually this is relatively slow and continuous but if the compressional force being applied suddenly increases, faulting may result, usually as a shear or thrust fault.

Figure 1.37: A small fold which has been thrust faulted, Redbank Plains, Queensland, Australia

Other types of folds include:

- **Isoclinal folds** are a series of several symmetrical folds aligned parallel to each other.

- **Slump folds** occur at the edge of sedimentary basins when an incompetent or weak bed is pushed down and along the margin of the basement of the basin by a heavier, more competent bed above. This may occur whilst the sedimentary beds are only in a moist, semi-solid phase of **lithification**.

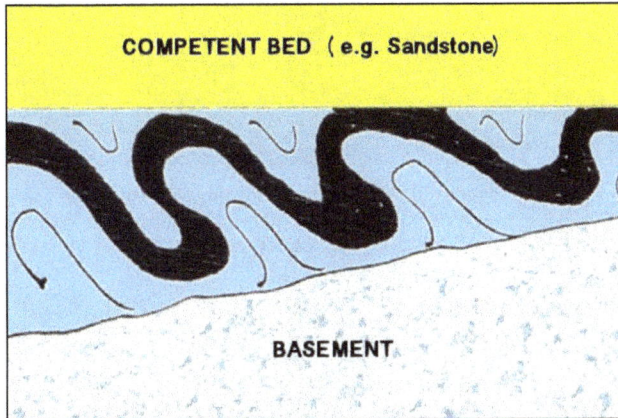

Figure 1.38: Diagram showing folding in a weak bed (e.g. shale) by a heavier competent bed (e.g. sandstone) above

Figure 1.39: Part of a slump fold of sandstones and shales (Evans Head, New South Wales, Australia)

Figure 1.40: Slump folding forming a structure called an unconformity (new sediments deposited following erosion of the fold), State Circle, Canberra, Australian Capital Territory.

- **Ptygmatic Folds** occur when originally flat and planar strata are bent or curved as a result of deformation, often by low-grade metamorphic action by compression. They may vary from microscopic to quite large in size.

Figure 1.41: A small ptygmatic fold of quartz veins within a hornfels rock

It is worth remembering that in reality, geological structures occur in 3-dimensions even though they are displayed in 2-dimensional photos and diagrams in textbooks. Most folds and faults taper off into the bigger geological scene, but some 3-dimensional folds like domes and basins, have been folded in several directions at once giving them their characteristic features.

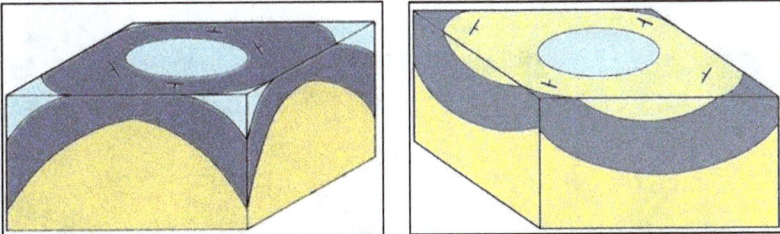

Figures 1.42 & 1.43: Block diagrams showing a dome (left) and a basin (right). Note the small map symbols showing the angle of tilt (dip) and direction of the bed (strike)) of the folded beds at those places.

Figure 1.44: Eroded and tilted limestone dome at Taemas, New South Wales, Australia

Figure 1.45: Wilpena Pound, in the Flinders' Ranges of South Australia is an eroded synclinal basin over 17 km. wide (Photo: Jo Blunn)

Chapter 2:   Volcanoes

2.1 Introduction

In some parts of the world, Earth movements and their subsequent events are more dramatic and occur in real time. Volcanoes have been known from ancient times, and many people live in regions where the activity of the local volcano has become a part of their normal lives. It is only when the volcano erupts that their lives are put in danger. Some volcanoes erupt without warning and sometimes as a complete surprise to the local inhabitants, who have not had any experience with volcanic activity. Both the disastrous eruptions of Mt. Vesuvius at Pompeii, Italy, in 79 A.D. and Mt. Lamington in Papua New Guinea in 1951, took the local inhabitants by surprise. The name volcano comes from the small volcanic island, Volcano in the Tyrrhenian Sea, about 25 km north of Sicily, thought by the Romans to be the home of their god, Vulcan.

2.2 Volcanic Eruptions

Volcanoes are now a well-known feature of many landscapes, even in countries where they are no longer active. Volcanoes in these places are **extinct** with no future activity probable, but in others, they may be **dormant**, showing no signs of activity but are likely to become active at any time. Volcanic activity at the Earth's surface can come in a great variety of forms depending upon the location and the nature of the rock type. Eruptions often occur suddenly, sometimes beginning as a large explosion of gas and rock material. The explosion may also be a sudden fountain of hot, molten lava. This has come to the surface from a pool of molten rock, or magma from below, and has erupted because of the increasing pressure of additional magma being produced in the **magma chamber** below the site of the volcano.

The composition of the magma will determine the type of volcanic eruption. Molten rock which has little or no quartz content, called **basic magma,** will cause relatively calm eruptions of fire fountains and fast-flowing lava with low viscosity as well as additional steam and other gases. Magmas with high silica content are called **acidic magmas**, and they can be very sudden and explosive, producing huge showers and rolling clouds of ash and rock called **pyroclastic flows** or **nuée ardente** (French: *glowing cloud*), as well as gases and slow-flowing, blocky lavas.

Pyroclastic eruptions produce igneous rocks having particles of different sizes. All pyroclastic particles, irrespective of size and shape are called **tephra:**

TEPHRA	SIZE	RESULTING ROCK
Blocks	>256 mm	Volcanic Agglomerate
Bombs	32 to 256 mm	
Lapilli	4.0 to 32 mm	Volcanic Breccia
Ash	0.062 to 4.0 mm	Tuff
Dust	<0.062 mm	

Table 2.1: Major pyroclastic tephra sizes

WIDTH = 30 cm.

Figure 2.1: A volcanic bomb - note the aerodynamic shape (Photo: Wikipedia)

In addition to solids, volcanoes also produce a very large amount of gas. This has been dissolved in the magma or has been produced as a result of interaction with surrounding water or rock. The most common gases from volcanic eruption are:

COMPONENT	PERCENTAGE
Water	77.0%
Carbon Dioxide	11.7%
Sulfur Dioxide	6.5%
Nitrogen	3.0%
Hydrogen	0.5%
Carbon Monoxide	0.5%
Sulfur Vapour	0.3%
Chlorine	0.05%
Argon	0.05%

Table 2.2: Major gases from volcanic eruptions

Poisonous methane gas (CH_4) may also come from the heating of vegetation by lava flows and other volcanic heat sources.

2.3 Shield Volcanoes

These are volcanoes which are large, have a rounded shape, and are almost entirely constructed from many layers of rapidly flowing lava which cools to form dark basalt or similar rock. The lava comes from a basic magma which has very little quartz content.

When it reaches the surface, it flows out relatively quickly and forms thick sheet-like lava flows which may cover an extensive area as **plateau basalts,** or may build up as many layers over a long period and form a large **shield volcano.** Sometimes the eruption may occur from long fissures, as in Iceland, or as a single fire fountain, throwing lava thousands of metres into the air such as from Kilauea in Hawai'i.

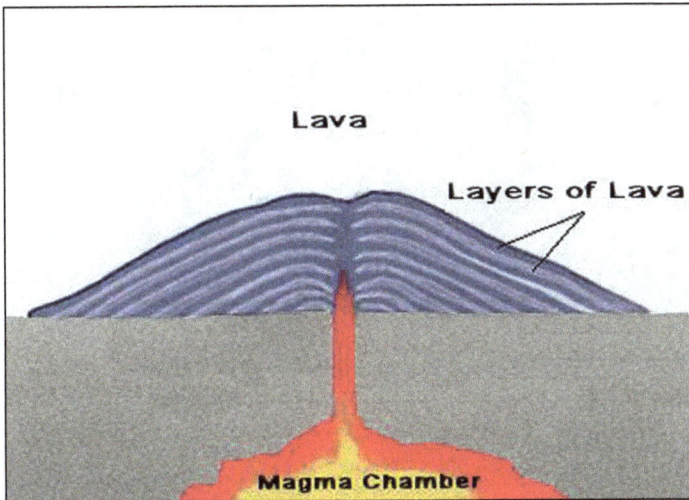

Figure 2.2: A diagram showing the structure of a shield volcano

Shield volcanoes are usually very large structures, such as those of the Hawaiian Island chain, which have been built up from the ocean floor about five kilometres below sea level. Mauna Loa, Mauna Kea and Kilauea on the island of Hawai'i are examples of such volcanoes. Currently, a new volcano, Lo'ihi is being built up, off Hawai'i's southeast coast, but it is still 1000 metres below sea level as a seamount. The Hawai'ian Islands are a good example of intraplate volcanoes which have formed within a tectonic plate. This is because they lie over a hot spot or **mantle plume**, a concentration of heat upwelling from deep within the mantle which re-melts the basalt of the crust above. Movement of the Pacific Plate to the northwest at Hawaii has resulted in the active volcanism appearing to move to the southeast where the current eruptions occur.

Figure 2.3: The rounded dome of Mauna Loa, Hawai'i (Photo: J.D. Griggs, USGS).

Figure 2.4: an old lava flow (from 1982) within the crater of Kilauea, Hawai'i

Figure 2.5: Lava flowing from a new vent in May 2018 from Kilauea volcano, Hawai'i during a series of massive eruptions (Photo: USGS)

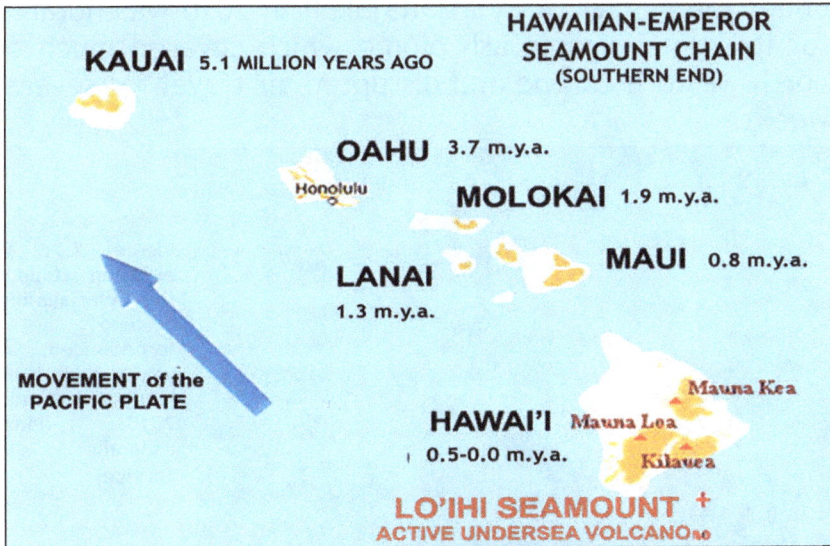

Figure 2.6: The Hawai'ian Islands showing the ages of volcanism in millions of years ago (m.y.a)

In Iceland, the volcanoes have similar eruptions of basaltic lava as individual volcanoes or as eruptions along a line or fissure. Iceland has been built up from the long-term eruptions at a mid-ocean ridge and continues to grow as the island moves apart from the ridge. Some of the volcanoes are particularly dangerous as they have formed below thick ice caps which also cover parts of Iceland. The sudden injection of lava into melting ice produces very large explosions of ash and large volumes of water which often flood parts of the island. These Icelandic ash clouds often go high into the upper atmosphere and travel long distances. In 1783-1784 the eruptions from the row of craters called *Lakagígar* (craters of Laki) produced an outpouring of ash and sulphur dioxide spreading a thick haze across Western Europe, resulting in many thousands of deaths throughout those two years. More recently, the eruption under the glacier of Eyjafjallajökull in 2010 was notable for its huge volcanic ash plume which covered much of north-western Europe and disrupted air travel for several weeks.

Figure 2.7: The eruption cloud of the Eyjafjallajökull volcano (pronounced: AY-yah-fyah-lah-YOH-kuul) in Iceland in 2010 (Photo: Icelandic Met. Service)

The volcanoes of the Galapagos Islands and Olympus Mons on the planet Mars are other examples of shield volcanoes.

2.4 Rhyolitic Domes

These come from the eruption of volcanoes from magma rich in silica and, are usually very are explosive. Because of this, they often look like non-volcanic, irregularly-shaped hills, and they usually explode without warning. When they produce lava, the rocks formed include rhyolite, dacite or trachyte, which are light in colour due to their high silica content. Often near the end of the first phase of the eruption, a solid dome or plug may be pushed up from the opening or vent. If the vent becomes blocked, pressure below may cause a secondary opening in the side of the volcano with hot, high pressure ash and gas being blown out with considerable force as a pyroclastic flow (nuée ardente), which may flow as fast as 700 km/hour and have a temperature of over 1000^{oC}. Mount Tarawera in New Zealand, Chaiten Dome in Chile, Mt. Pelee in Martinique and Lassen Peak in the USA are examples of such violent, unpredictable volcanoes.

These volcanoes are extremely dangerous because of their sudden and violent nature. The unpredictability of eruptions of nuée ardentes is particularly dangerous, because of their rapid flow down valleys and into areas of habitation.

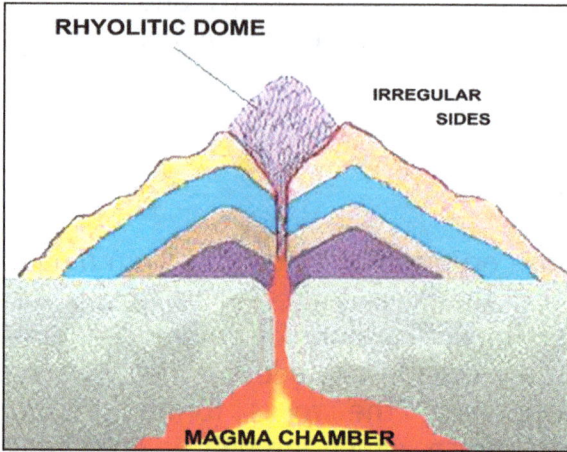

Figure 2.8: A diagram showing the structure of a dome volcano

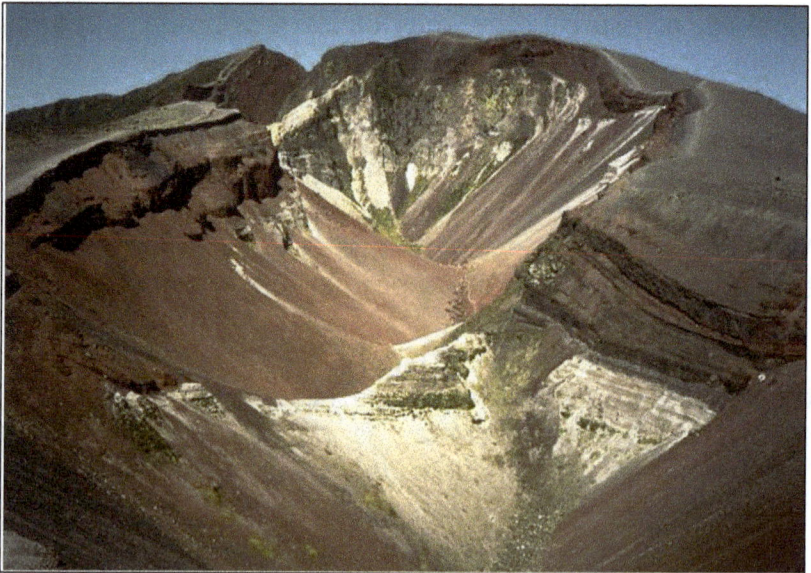

Figure 2.9: The irregularly-shaped Mt. Tarawera of North Island, New Zealand which erupted violently without warning in 1886 causing much destruction and killing 158 people.

2.5 Cinder Cones

These can form as individual volcanoes or as **parasitic cones,** which are small cones built up on the sides of larger volcanoes by secondary eruptions. They are built up by the eruption of fine ash or larger, rougher, blocky fragments called **scoria.** They can occur suddenly as did Volcan de Paricutin, in Mexico which suddenly grew out of a flat cornfield into a steep-sided cone in 1943. Cinder cones can be a few metres high when they are on the flanks of older volcanoes or to over 400 metres high in the case of Paricutin. Cerro Negro in Nicaragua is perhaps the most active of such volcanoes.

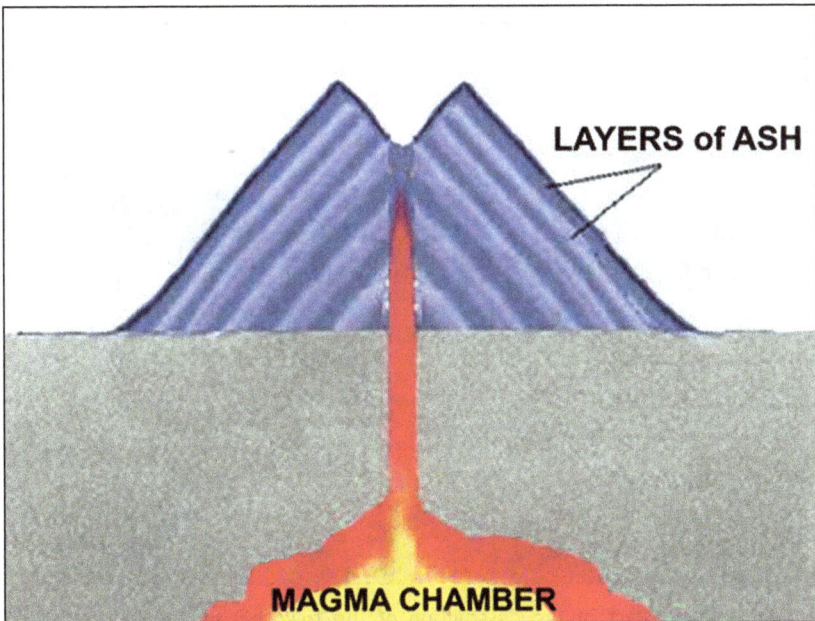

LAYERS of ASH

MAGMA CHAMBER

Figure 2.10: A diagram of a cinder cone

Figure 2.12: Part of the destruction of the township of Rabaul, New Britain, Papua New Guinea by the eruption of Tavurvur Volcano in 1994 (Photo: Karne Povey)

Figure 2.13: Looking in the same direction towards active Mount Tavurvur, Rabaul, Papua New Guinea today.

2.6 Stratovolcanoes

The most common type of volcano, however, is the **stratovolcano** or composite volcano, which is made up from alternative layers of both ash and lava. These give the uniform slopes of the classic volcanic cone, and are the most well-known of all volcanoes. The eruption usually begins explosively, often with little warning, with a large amount of hot ash and gas going to great heights. This can then collapse with the ash flowing rapidly over the surface as a pyroclastic flow.

Perhaps the most famous eruption of this type was that of Mount Vesuvius in August, 79 AD, described by **Pliny the Younger** (Roman: 61-131 AD), an eye-witness, who described the actions of his uncle, the admiral and naturalist, **Pliny the Elder** (Roman: 23-79 AD), who attempted to save the people of the city of Stabiae, near Pompeii, which was destroyed by the eruption. Such a violent eruption is now termed a Plinian Eruption.

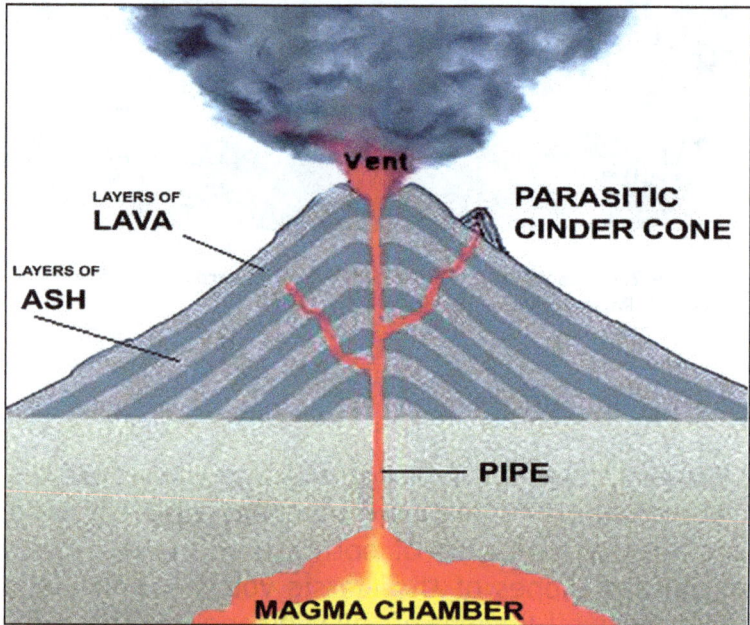

Figure 2.14: A diagram showing the structure of a stratovolcano

Figure 2.15: Mount Vesuvius looms over the Forum of Pompeii, now a suburb of the Italian city of Naples, only a few kilometres from the volcano

Figure 2.16: A plaster cast of one of the victims of the 79 A.D. eruption at Pompeii. The cast was made by pouring plaster into the cavity left in the thick ash by the body after the soft tissue had disintegrated over time.

Online Video 2.1: Explore the ruined city of Pompeii, Naples, Italy. Go to https://youtu.be/PevacwLMvWU

Plinian eruptions are very destructive and have a **Volcanic Explosivity Index** (VEI) of between 4 and 6. The VEI is an open ended scale of the explosive nature of a volcano. It was devised in 1982 by members of the United States Geological Survey (USGS) and the University of Hawai'i and its scale ranges from a value of 0 is for non-explosive eruptions to that of the largest volcanoes in history given a magnitude 8. These arbitrary values have been defined as less than $10,000 \text{ m}^3$ of tephra ejected for the 0 values and 8 for an extremely large explosive eruption ejecting more than $1.0 \times 10^{12} \text{ m}^3$ of tephra.

Volcanic ash clouds can be extremely dangerous because they can cover a wide area very quickly. Locally, the sudden and very dramatic fall of huge volumes of ash leads to suffocation of those who cannot escape, collapse of roofs from the weight of ash, destruction of any machinery which takes the ash in through air intakes, the loss of crops which are covered with ash, and the loss of stock eating plants contaminated with volcanic dust containing compounds such as hydrogen fluoride.

On a more global scale, very fine clouds of ash are able to be dispersed over very large areas because they are ejected to high altitudes where prevalent strong winds can carry the ash for thousands of kilometres. The eruption of Mount St. Helens in 1980, like many of these ash eruptions, was **phreatic**, that is, it was extremely explosive because of the amount of steam generated by the eruption. This sent gases and fine ash well to the southeast across much of the United States, producing a huge ash cloud which covered much of central, northern states. Ash from the Indonesian volcano Galunggung, almost brought down a British Airways 747 in 1982, and the more recent ash eruption of the Icelandic volcano Eyjafjallajökull in 2010, though relatively small for volcanic eruptions, caused enormous disruption to air travel to and from Great Britain and northern Europe. Apart from personal danger and loss to the local inhabitants, ash eruptions can have a considerable negative effect on the global economy. There have also been hypotheses that largescale ash eruptions have been responsible for some of the extinctions of species on Earth due to the great reduction of sunlight which has been obscured by ash covering most of the globe. Often the Plinian phase is followed by lava flows of a slow, blocky rhyolitic nature which moves down local valleys, destroying everything in its path. After the eruption is over, and the magma chamber below has been depleted, the top of the volcano may collapse and form a large, near circular depression called a **caldera** from the Spanish for a cauldron.

Other stratovolcanoes include Mount Taranaki in New Zealand, Mount Fuji in Japan, Mount Mayon in the Philippines, Mount Nyiragongo in Africa, and the Nevada del Ruiz volcano in Columbia. This latter volcano erupted suddenly in 1985, producing a very large, fast-flowing stream of hot, quick-setting mud which was formed by a mixture of the hot volcanic ash and meltwater from the volcano's snowcap. Such a mudflow is called a **lahar**, and in the Nevada del Ruiz eruption, such a flow swept through the town of Armero killing 25,000 people.

The Pacific Ocean contains, and is ringed by many active volcanoes, including those of New Zealand (Tarawerra, Taranaki, White Island/Wakaari, Ngaurahoe, Tongariro, Ruapehu to name a few), Tonga (Hunga Tonga-Hunga Ha'apai), Vanuatu (Yasur, Ambryn), Papua New Guinea (Tavurvur, Lihir), Japan (Sakurajima, Asama), the Kamchatka Peninsula of Russia (Bezymianny, Shivluch), Alaska and the American west coast (Pavlof, Katmai, St. Helens, Lassen Peak), Hawaii (Mauna Loa, Mauna Kea, Lo'ihi Seamount, Kilauea), Central and South America (Arenal, Cerro Negro, Cotopaxi, Lascar, Tungurahua, el Misti) and the Antarctica Peninsular (Deception Island).

Figure 2.17: The violent eruption of Mount St. Helens, 1980 (Photo: USGS)

The location of many volcanoes surrounding the Pacific Ocean gives it the name of the Pacific Ring of Fire.

Figure 2.18: Mount Taranaki on the North Island of New Zealand

Figure 2.19 Yasur, the active volcano on Tana Island, Vanuatu, South Pacific just before it erupted in 1999.

Online Video 2.2: Travel to Yasur, the active volcano of Vanuatu in the south Pacific and drive over the moon-like landscape. Go to https://youtu.be/oML8kdnfLYU

Figure 2.20: Approaching Deception Island, just off the coast of the Antarctic Peninsula. Its caldera has exploded leaving a horseshoe-shaped lagoon in the centre which is open to the sea.

Figure 2.21: Inside the lagoon can be seen the remains of the old whaling Station, Deception Island, destroyed in the 1970 eruption

Online Video 2.3: Sail into the caldera of Deception
Island, Antarctica - a dormant stratovolcano
Go to https://youtu.be/Y0z2MLQ391U

Sometimes there are undersea volcanoes, especially in the Pacific and north Atlantic oceans. Here new islands can be formed such as Surtsey, named after *Surtr*, a fire giant from Norse mythology, just south of Iceland in 1963, and Hunga Tonga in the Pacific near Tonga in 2015. Large rafts of pumice stone can be formed from such undersea eruptions. Pumice is a rock which is very vesicular due to the original frothing at the top of felsic lava and is less dense than water, so it floats. In August 2012, a large pumice raft appeared near New Zealand which was over 480 km long, 48 km wide, and sixty centimetres thick. Such rafts are a hazard to shipping by fouling water intakes and propellers and even impede the motion of smaller craft.

2.7 Other Volcanic Hazards

There are often other forms of volcanic activity near the main volcanic site which are hazardous even if the volcanoes themselves are dormant. These may include hot springs and mud pools, **geysers**, **fumaroles** (gas vents) and **solfataras** (sulfur vapour fumaroles).

Figure 2.22: The Pohutu Geyser, Rotorua, New Zealand

Online Video 2.4: Visit the geysers and mud pools of the thermal area at Rotorua, North Island, New Zealand
Go to https://youtu.be/0MHqzOQpWWY

Figure 2.23: Hot mud pool, Rotorua, New Zealand

Figure 2.24: A solfatara or sulfur gas vent in the inner crater of Kilauea, Hawai'i

Online Video 2.5: Visit the solfatara in the inner crater of Kilauea, Hawai'i
Go to https://youtu.be/Or2NVqDBlbo

In places where lava has regularly flowed over the land surface, the ground below may be honeycombed with lava tubes or caves. These have been formed when fast-flowing lava, such as basalt, runs over the surface in a long tongue-like structure and then cools quickly, hardening on the outside. The hotter inner core continues to flow, leaving a hollow tube inside. The roof of these lava tubes is often very thin and sometimes an end will collapse, giving access to the tube. They also constitute a hazard for those who may walk over the lava field as the roof of any cavity below may collapse.

Figure 2.25: Entrance to a lava tube cave behind a waterfall near Hilo, Hawai'i

Figure 2.26: Inside the Thurston Lava Tube, Hawai'i

Online Video 2.6: Walk underground through the Thurston Lava Tube, Hawai'i, USA
Go to https://youtu.be/JM9usl-I6SU

Of special interest are the so-called super volcanoes, usually referred to by their resulting **mega calderas** by vulcanologists. These exist as wide calderas, often filled with water forming large lakes. These once erupted with great power, giving ejecta masses as very large volumes of pyroclastics, of over 10^{15} kilograms. They exist as dormant regions, but it is thought that they lie over unstable hot spots in the Earth's crust. These are zones of excessive heat flow from the Earth and are caused by large pools of magma which has risen from the mantle below the crust and now are trapped in the crust above. Yellowstone Caldera in the USA, Lake Taupo in New Zealand and Lake Toba in Indonesia may have been sites of super volcanoes.

The secondary effects of a volcanic eruption can also be fatal. These are environmental and social effects which occur as a result of the eruption include destruction of crops and livestock by ash fall. Stock often die from eating ash-contaminated vegetation or directly from the effects of gases such as carbon dioxide and sulfur dioxide. Water courses which were vital to both stock and humans can become contaminated with the injection of volcanic gases and solutions and also because of any damming effects which may occur upstream due to ash falls or lava flows cutting across the water source.

Ash falls can occur locally but also on a global scale as an immense eruption of ash can be carried by the winds of the upper atmosphere to distant parts of the world. For example, the eruption of the Huaynaputina volcano in Peru in 1601 caused a massive famine in Russia which killed over two million people.

Starvation and disease, destruction of villages, seaports and roads have been a feature of many volcanic eruptions in historical times. About 1627 B.C. the stratovolcano of Thera (modern Santorini, Greece) exploded killing many on the island itself but also destroyed the Minoan civilization on the island of Crete 100 kilometres to the south.

Figure 2.27 Part of the remains of the island of Thera (Santorini) today, seen from the current dormant volcanic island of Nea Kameni, which has grown within the old caldera which formed a large lagoon after the explosion

Figure 2.28 Remains of the ancient city of Akrotiri on Santorini, Greece. It was part of the Minoan civilization but was suddenly buried when the island of Thera exploded. Some archaeological evidence suggested that there may have been an initial quiet period during which some people returned before the major eruption

Online Video 2.7: Sail to Neo Kameni, the dormant volcano in the ocean caldera of Santorini (Thera), Greece
Go to https://youtu.be/EaOP0Q4TYUg

Year	Volcano (and country)	Fatalities
1815	Tambora (Indonesia)	92,000
1883	Krakatau (Indonesia)	36,417
1902	Mount Pelee (Martinique)	29,025
1985	Ruiz (Colombia)	25,000
1792	Unzen (Japan)	14,300
1788	Laki (Iceland)	9,350
1919	Kelut (Indonesia)	5,110
1882	Galunggung (Indonesia)	4,011
1631	Vesuvius (Italy)	3,500
79	Vesuvius (Italy)	3,360

Table 2.3 Some of the most destructive volcanoes with local fatalities
(Data from *Volcanic Hazards: A Sourcebook on the Effects of Eruptions* by Russell J. Blong Academic Press, 1984).

2.8 Major Active Volcanoes Today

There are about 400 to 600 active volcanoes around the world. Some of the most violent volcanoes are located in a large circle around the Pacific Ocean called The Ring of Fire. These volcanoes produce ash and lava of andesitic composition and so this ring is also termed the Andesite Line. Some of the most common volcanic areas are shown on the map below (numbers refer to volcanoes or groups of volcanoes):

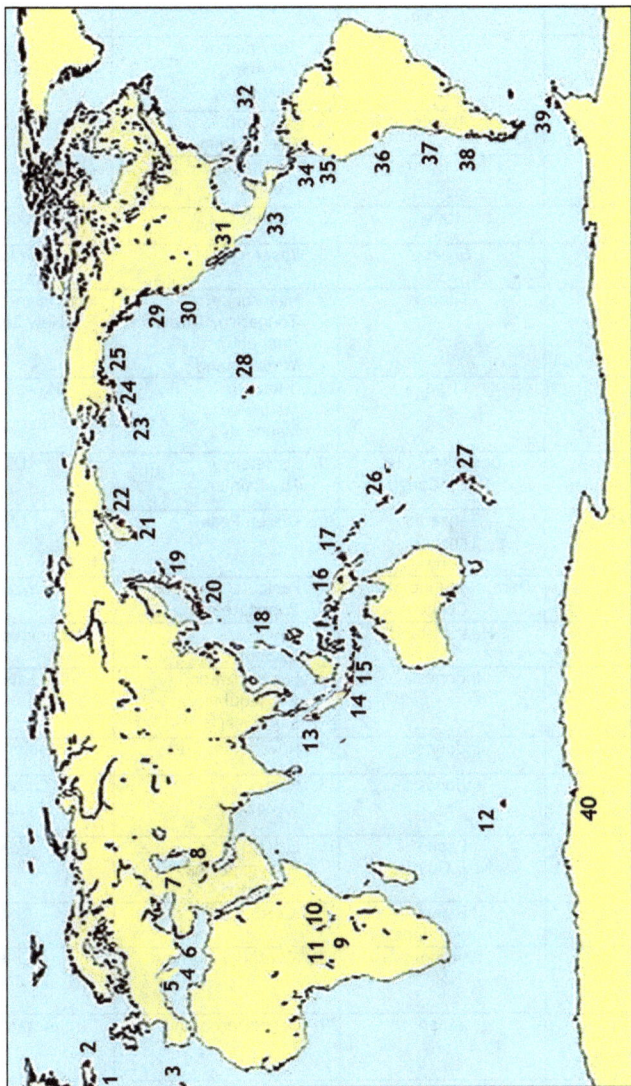

Figure 2.28: A Map showing the locations of some of the world's act volcanoes (refer to the chart next page)

Volcano or Group	Location	Volcano or Group	Location
1. Laki, Hekla, Surtsey, Eyjafjallajökull	Iceland	21. Alaid	Kiril Islands, Russia
2. Prestahnúkur, Krafla,	Iceland	22. Bezymianny, Karymeky, Mutnovsky	Kamchatka Peninsula, Russia
3. Faial, Pico	Azores	23. Bogoslof, Great Sitkin	Aleutian Islands, USA
4. Volcano, Stromboli Etna	Lipari Is., Italy	24. Shishadin, Pavlov	Aleutian Islands, USA
5. Vesuvius	Italy	25. Katmai	Alaska, USA
6. Thera (Santorini)	Greece	26. Yasur	Tana Island, Vanuatu
7. Mount Ararat	Turkey	27. Ngauruhoe, Tongariro, Tarawera, Ruapehu, White Island	North island, New Zealand
8. Demavand	Iran	28. Kilauea, Mauna Loa, Mauna Kea	Hawaii, USA
9. Nyiragongo	Dem. Republic of the Congo	29. St Helens Mt. Rainier	USA
10. Ol Doinyo Lengai, Kilimanjaro, Mt. Kenya	Tanzania Tanzania Kenya	30. Lassen Peak	USA
11. Nyamuragira	Dem. Republic of the Congo	31. Paricutin, Jorillo, Popocatepetl	Mexico
13. Big Ben	Heard Island	32. Pelee La Soufrere	Martinique, France
13. Merapi Dempo	Indonesia	33. Fuego, Izalco, Consequina, Santa Maria	El Salvador
15. Krakatau	Indonesia	34. Nevado del Ruiz	Colombia
15. Tambora Agung	Indonesia	35. Purace Cotopaxi	Colombia Ecuador
16. Lamington	Papua New Guinea	36. El Misti, Huaynaputina, Guallatiri	Peru
17. Karkar	Papua New Guinea	37. Cerro Azul	Chile
18. Mayon, Taal Pinaturbo	Philipines	38. Osorno & Burney	Chile
19. Fujiyama, Unzen Bandaison	Japan	39. Deception Island	Antarctica
20. Myozin-syo	Japan	40. Erebus, Terror	Antarctica

Table 2.4: Locations of some of the world's active volcanoes

2.9 Extra-terrestrial Volcanoes

Volcanoes exist on other planets and some of their moons within our solar system. Like Earth, Venus and Mars are believed to have hot interiors and are continuing to lose heat. Surface effects of tectonism are due to convection currents bringing heat to the surface to produce volcanoes. Whilst Mars appears to have a primitive form of plate tectonics, Venus does not show plate development although it does have surface rifts and volcanoes.

Mercury and the Moon both have extensive, flattened lava plains suggesting volcanic activity in the distant past but there is no evidence of current volcanism.

Mars is smaller than Earth and has cooled more, producing a thick outermost layer. The convection activity on this planet appears to be restricted to a

Figure 2.29: Some of the volcanoes on Mars. The largest, Olympus Mons is seen at top left and below it is a line of smaller volcanoes: Arsia Mons (bottom), Pavonis Mons (center), and Ascraeus Mons (top)

few locations where hot material may be rising from the interior toward the surface, causing the surface to bulge, stretch, and crack. The largest of these areas is the Tharsis Bulge and the Valles Marineris, a large rift, where the surface has split apart. The thickness of the surface layers and the reduced gravitational field has allowed the formation of the Solar System's largest volcano – Olympus Mons

Venus also shows evidence of tectonic activity, where the surface has been, in some locations, stretched and broken, and in other regions, crumpled. Most of the planet's surface consists of lava plains dotted with many large

Figure 2.30: Maat Mons, one of the volcanoes of Venus.

isolated shield volcanoes and huge, ring-shaped structures hundreds of across. These rise hundreds of meters above the surface and is believed that they are formed when plumes of rising hot material in the mantle push the crust upwards into a dome shape and then collapse in the centre leaving a crown-like structure called a corona. Volcanoes also seem to be of the shield type.

Volcanism in the outer planets is uncertain because of their thick, dense cloud covering, however Mercury may have had volcanism a very long time ago to produce lava flows seen within some craters by NASA's *Messenger* spacecraft.

Several of the moons of the outer planets do show current surface tectonism. Io, the innermost of Jupiter's four largest moons, is perhaps the most active object in the Solar System. This is due to its internal heating which is caused by the tidal forces of its parent planet, Jupiter, trying to pull the moon apart. Many mushroom-shaped plumes of sulfur and solid sulfur dioxide have been observed shooting out from the moon's surface and lava plains seem to consist of these materials as well as some silicates (minerals such as orthopyroxene containing iron, magnesium as well as silicon and oxygen).

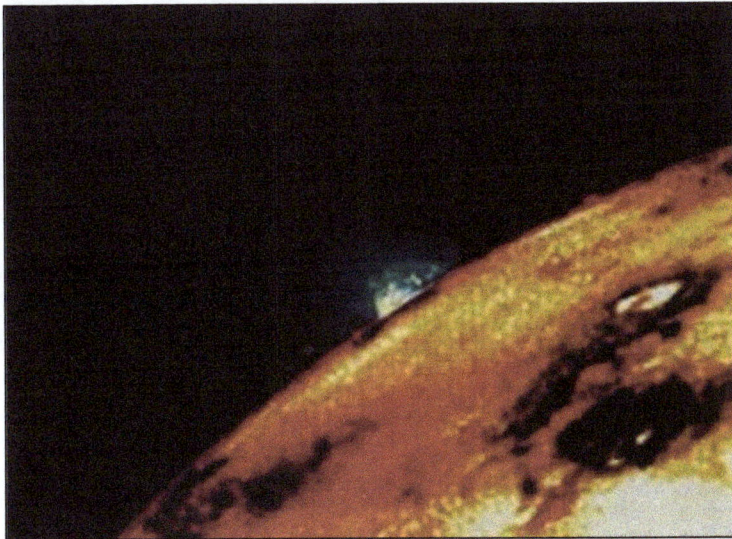

Figure 2.31: A sulfur plume erupting from the surface of Jupiter's moon Io

Other moons in the Solar System also experience tectonism due to tidal (or gravitational) forces due to their parent planet. These often show **cryovolcanism** – that is, cold volcanic eruptions of gases such as nitrogen and liquids such as water from their surfaces. The ice covered moons Triton (of planet Neptune) and Enceladus (of the planet Saturn) both show extensive plumes or geysers of gas and water from their surface, suggesting that the heated water below the ice covering may have conditions suitable for primitive aquatic life. It is also possible that other ice moons such as Europa, Titan, Dione, Ganymede, and Miranda also have had similar activity.

Figure 2.32: Water and gas geysers erupting from the ice moon of Saturn, Enceladus in 2017

2.10 Volcano Emergency Plan

People who live within the range of a volcano are well aware of the dangers they may have to face one day. However, where they live is their home; it is where their people have lived, and the soil is often rich because of previous eruptions. Besides, the volcano is not erupting today.

Figure 2.33: Volcán Tungurahua (Chechua language: Throat of Fire) overlooking the town of Baños de Agua Santa, Ecuador last erupted in March, 2016

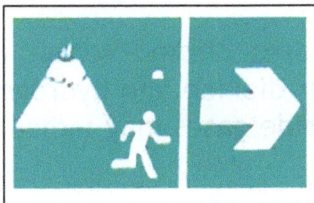

Figure 2.34: An emergency sign in the town of Baños de Agua Santa, Ecuador on the slopes of Volcán Tungurahua

73

Like many natural disasters, volcanic eruptions sometimes come at unpredictable times. Seismologists have a wide range of instruments and knowledge about volcanoes, and most countries with active volcanoes have them monitored at all times. Possible signs of an imminent eruption can be detected by sudden changes in activity in and around the volcano. These may include:

- Seismic tremors increasing in size and frequency. These may be detected only by the seismographs embedded on the volcano.

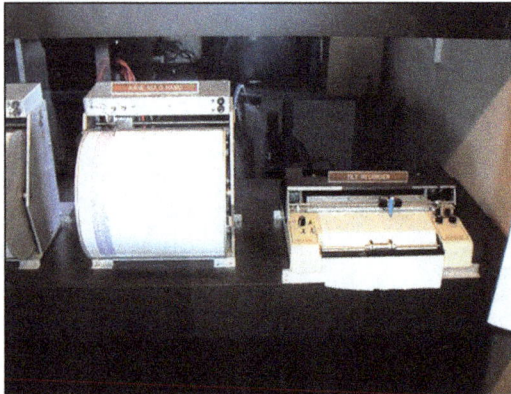

Figure 2.35: An early seismometer near the crater of Kilauea, Hawai'i

- Change in temperature or colour of any volcanic-fed spring, pool or stream.

- Rise in water level in local boreholes and wells as subsurface pressures from new magma injection may increase the height of the water table.

- Release of additional gases from vents or fumaroles. Increase in sulfur dioxide and hydrogen sulphide and hydrogen chloride gases may indicate an imminent eruption.

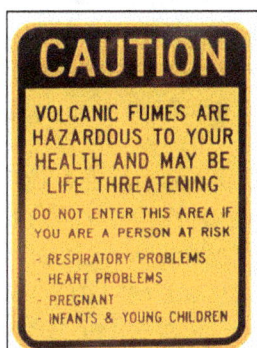

Figure 2.36: Sign inside the crater of Kilauea, Hawai'i

- Change in shape of the volcano's sides. This can be measured by **tiltmeters** which are instruments which can measure very slight changes in the orientation of the surface of the volcano using a LASER beam. Changes in the volcano's surface usually occur because of the pressure of new magma coming into the magma chamber below the volcano.

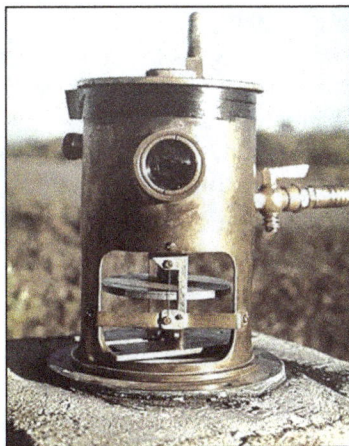

Figure 2.37: An early tiltmeter embedded into the rock on the slopes of Mauna Loa, Hawai'i (Photo: USGS)

- Sudden changes in animal behaviour can sometimes provide a possible warning signs because animals are more sensitive to the ground vibrations and release of gases.

Most countries with active volcanoes have a colour-coded system of hazard alerts. For example, on Tana Island, Vanuatu is the active volcano Yasur, which under normal conditions will send a smoke cloud or ring into the sky about every 15 minutes. In 1999, increased activity in the lava pool in the vent suggested that an eruption of this explosive andesitic ash volcano could occur and so the alert status went up to Stage 4:

LEVEL	STATUS
0	Normal low-level activity
1	Increased activity, danger near crater only
2	Moderate eruptions, danger close to the volcano vent, within parts of Volcanic Hazards Map Red Zone
3	Large eruption, danger in specific areas within parts of Volcanic Hazards Map Red and Yellow Zones
4	Very large eruption, island-wide danger (including areas within Red, Yellow and Green Zones)

Table 2.5: Alert status code for Vanuatu, southern Pacific

Figure 2.38: The author in front of Yasur volcano, Tana Island, Vanuatu in 1999 with the volcano at Level 4 alert just before a major ash eruption (see Video 2.2)

If a volcano is likely to erupt nearby, it is prudent to remember the nature of the main hazards associated with eruptions. These include:

- Eruption of pyroclastics varying from large, semi-molten bombs near the volcano to very fine ash at distances up to many kilometres away. The fine ash is able to penetrate closed doors and covered machinery and is abrasive to motors and lungs. Death by suffocation usually occurs before the weight of ash collapses any roof structure. If escape is impossible, staying indoors and masking the face with a wet towel is the best option.

77

- Poisonous gases, especially sulfur dioxide and hydrogen sulfide gases which are choking, and carbon dioxide and carbon monoxide which are odourless, will all cause death quickly if inhaled. This hazard is particularly dangerous in some of the African volcanoes which have large crater lakes, such as Lake Nyos in the Cameroons. In 1986, a dense cloud of heavy carbon dioxide was emitted from the crater's lake and rolled down the valleys into low-lying areas killing over 1700 people and all other animal life. This was an insidious event, as the gas is invisible and odourless and quickly kills by preventing oxygen intake. In the case of Lake Nyos, the gas came from an over-turning of the lake floor which contained considerable amounts of organic matter. Such an eruption is called a **limnic eruption.** If gases are detected in large amounts, evacuation is the only choice, using a wet towel as a mask and avoiding any low areas which may contain pockets of dense gas.

- Nuée ardentes are fiery clouds of pyroclastics and gas which explode out sideways from a vent and roll down local valleys. These usually have temperatures of several hundred degrees Celsius and move with velocities of several hundred kilometres per hour. Prior evacuation is really the only defence for these destructive flows;

- Lava flows are more avoidable except near shield volcanoes such as in Hawai'i. Here, the rapid flow of some lavas close to the vent are capable of trapping the unwary who venture into valleys down which the lava could flow. During some eruptions in Iceland and Italy, the locals have had some success in channelling lava flows away from places of habitation using high pressure hoses and constructed channels; and

- Lahars or mudflows may also come down the valleys from a distant, snow-capped volcano which has increased its vent temperature as it has erupted. They can flow at great speeds for long distances so the actual volcanic event which triggered the flow may not have been detected. Many of the volcanoes in the Americas are potential lahar hazards. These flows can only be avoided by moving to high ground or complete evacuation.

If a local eruption has occurred, the type of response will depend upon the nature of the eruption (ash, lava or both) and may involve:

- Monitoring media reports which are usually given at regular intervals. It is useful to have a radio which uses batteries in case the local power fails.

- Evacuation as directed by civil emergency or local authorities.

- Seeking protection from gases or falling ash by staying indoors if evacuation is not immediately possible. Use a wet towel as a mask if breathing is difficult and also bring animals and pets indoors as well.

- Have extra supplies of water and food as the water supply may not be available because of disruption or contamination. Food supplies such as local crops and shops may be destroyed by the eruption. An emergency kit would be a good precaution in places known for regular eruptions. This should also include a small shelter and additional protective clothing.

- Refraining from driving or flying during ash eruptions because visibility and traction will be difficult and ash will soon damage and stop any engine.

- Avoidance of places where structural damage by the weight of ash, lava flow or earth tremor may cause a roof or building to collapse. Fallen power lines would also be a hazard.

Whilst an area may be well-monitored, the nature and size of a sudden eruption can be overwhelming. Whilst an alert may be premature, it is wise to consider evacuation to a place on higher ground, upwind and at some distance from the volcano. The aftermath of the eruption may be a trial for local communities because of the lack of water, food and shelter and health issues due to contamination of water and lack of waste disposal. Damage to property and livelihood may also be longer term and a blow to any society's economy.

Chapter 3:   Earthquakes

3.1 Introduction

Some movements on the surface of the Earth can be unexpected and sudden, causing massive destruction and loss of life in an instant. In many places around the world, the movement of individual parts of the Earth's crust, called **tectonic plates,** can cause very large pressures to build up within the crust. When this pressure overcomes any frictional forces, or the stress exceeds the strain of the local rock, there is a sudden movement or breakage. This movement at plate boundaries or along faults at plate edges and within plates can cause small tremors or larger earthquakes. When the internal pressure builds up along the fault or boundary, the rock deforms rock until it suddenly breaks and rebounds. This releases a tremendous amount of energy which spreads out rapidly in all directions as a series of shock waves. The concept of how this happens is called the **elastic rebound theory.**

3.2 Location of Earthquakes

The shock of this movement or break is transmitted as waves which start at the place where the earthquake waves are produced called the **focus** or **hypocentre** from the Greek *hypókentron* for below the centre.

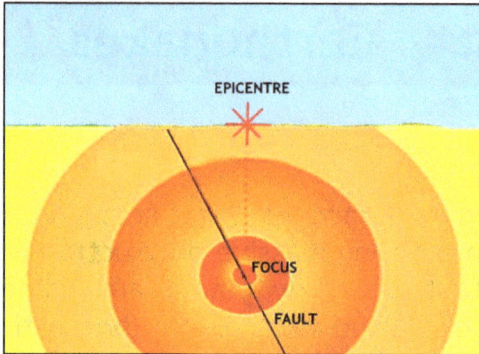

Figure 3.1: Focus and epicentre of an earthquake

This usually occurs well below the surface of the Earth at various depths. Normally, deeper earthquakes are the most energetic because of the greater weight of the moving rock above it and therefore the greater energy which will be produced. These waves travel through and around the Earth as a result of the sudden release of strain and the place on the Earth's surface, immediately above the focus where the maximum energy is felt is called the **epicentre**, from the Greek *epikentros*, meaning situated on a centre.

Earthquakes occur very frequently with many hundreds of small tremors occurring each day in some earthquake-prone centres. Most of these go unnoticed, however occasionally when the energy builds up for an extended time and is suddenly released, a major earthquake will cause considerable damage.

Earthquakes have been recorded almost as long as humankind has kept records. Some of the biggest earthquakes recorded are shown in the table on the next page:

YEAR	REGION	MAGNITUDE	FATALITIES
856	Damghan, Iran	unknown	200,000
1556	Shansi, China	unknown	800,000
1737	Calcutta, India	unknown	300,000
1755	Lisbon, Portugal	8.7	70,000
1812	New Madrid, USA	7.9	few
1906	San Francisco, USA	7.7	3000
1920	Gansu, China	8.6	200,000
1932	Gansu, China	7.6	70,000
1933	Sanriku, Japan	8.4	3000+
1960	Southern Chile	9.5	6000
1970	Northern Peru	7.7	85,000
1976	Tangshan, China	8.5	250,000
1988	NW Armenia	6.8	55,000

Table 3.1: Some of the largest earthquakes (Data from USGS)

Figure 3.2: Major damage in Turkey, 1999 (Photo: USGS)

Figure 3.3: Major damage from the Sichuan, China, earthquake of 2008 (Photo: USGS)

There have also been several destructive earthquakes in modern times:

YEAR	DATE	REGION	MAGNITUDE	FATALITIES
1990	20 June	Iran	7.4	50,000
1991	19 October	Northern India	6.8	2000
1992	12 December	Flores, Indonesia	7.8	2519
1993	29 September	India	6.2	9748
1994	6 June	Colombia	6.8	795
1995	16 January	Kobe, Japan	6.9	5530
1996	3 February	Yunnan, China	6.6	322
1996	17 February	Irian Jaya, Indonesia	8.2	166
1997	10 May	Northern Iran	7.3	1572
1998	30 May	Afghanistan Border	6.6	4000
1999	17 August	Turkey	7.6	17118
1999	20 September	Taiwan	7.7	2297
2000	4 June	Southern Sumatra	7.9	103
2001	26 January	India	7.7	20023
2001	23 June	Coastal Peru	8.4	138
2002	25 March	Hindu Kush Region	6.1	1000
2003	26 November	Southeastern Iran	6.6	31000
2004	26 December	Northern Sumatra	9.1	227898
2005	28 March	Northern Sumatra	8.6	133
2006	26 May	Java, Indonesia	6.3	5749
2007	15 August	Central Peru	8.0	514
2008	12 May	Sichuan, China	7.9	87587
2009	29 September	Samoa	8.1	192
2009	30 September	Southern Sumatra	7.9	1117
2010	12 January	Haiti	7.0	316000
2010	27 February	Maule, Chile	8.8	507
2011	11 March	Honshu, Japan	9.0	20896
2012	6 February	Philippines	6.7	113
2013	24 September	Pakistan	7.7	825
2014	3 August	Wenping, China	6.2	729
2015	25 April	Nepal	7.8	8964
2016	24 August	Italy	6.2	298

Table 3.2: Some of the most recent major earthquakes (Data: USGS)

3.3 Earthquakes Effects

Apart from structural damage, other effects of earthquakes are:

- **Tsunamis** are wrongly termed tidal waves and have little to do with the tidal pull of the Moon and Sun. They are produced when there has been a large <u>vertical</u> movement at a plate boundary or fault below the ocean. They can also be caused by explosive volcanic eruption or landslides in confined waterways such as the 914-metre-high wave generated in Lituya Bay, Alaska in 1958.

  The sudden vertical motion causes the sea to generate an ocean wave of large wavelength and **amplitude** or wave height, which builds up as a vertical wall from a few centimetres to several metres in height when it reaches the shallows of the coastline. This wave often appears on the horizon as a very long, straight wall of water, and whilst some of them may be only a metre or so in height, they come onto land with the entire weight of the ocean behind them. This is unlike the normal wind-driven surfing wave which has very little water behind it. This can be very deceptive when the tsunami looks to be only knee deep, but even at that height it will sweep away people, trees and many buildings.

Figure 3.4: Tsunami damage in Banda Aceh, Indonesia 2004
(Photo: Guy Gelfenbaum USGS)

- **Seiches** are standing waves which go to-and-fro within smaller bodies of water such as lakes and pools, and are generated by the seismic waves passing through them.

- **Sand geysers** are produced by the forces within the ground compressing the soil and causing it to erupt through the surface as a semi-liquid geyser. Some clay, called **thixotropic** clay, which is normally very solid, will suddenly liquefy when shaken. When this occurs, any structure built upon such clay will collapse. This **liquefaction** of the ground will often cause the collapse of buildings and the breakup of roads and other surface structures.

- **Mole tracks** are localised ground disturbances consisting of small, interrupted uplifts and small to large escarpments due to compressional forces.

Figure 3.5: Mole track disturbances from an earthquake at Izmit, Turkey. (Photo: USGS)

- **Electrical fireballs** which are sometimes seen above the ground as the electrical charges within the minerals and solutions as ions in the ground are unbalanced and form excessive electrical charge above ground.

- **Landslides** are often triggered in many of the mountainous areas prone to earthquakes, especially in places such as in the Himalayan regions of Afghanistan and Pakistan and other mountainous areas. These often prove to be the most catastrophic event, covering whole villages, damming rivers and cutting roads and other communication networks.

Figure 3.6: Landslide near Hongyang, China (Photo: USGS)

Whilst most of the world's biggest earthquakes usually occur along tectonic plate margins, such as subduction zones and transform faults, earthquakes of significant sizes can occur along older fault lines within plates. These earthquakes occur when the huge forces caused by plate movement are transferred through the plate, building up pressure along old fault lines to the point where the fault will suddenly move or rock layers break.

In Australia, a country noted for its lack of seismic activity, there have been moderate earthquakes at: Meckering, Western Australia, which occurred on 14th October, 1968, along the Darling Fault (Richter 6.9, 20 injuries, some damage); Tennant Creek, Northern Territory, (at 6.7 it caused minor damage); and at Newcastle, New South Wales, 28th December, 1989, (at 5.6 it caused 13 fatalities and 130 injured). Such earthquakes are called **intraplate earthquakes.**

3.4 Detecting Earthquakes

Earthquake waves are detected by **seismographs,** also called seismometers, which are instruments which make use of the resistance to movement or **inertia,** of large masses. One of the first devices constructed specifically to locate earthquakes, was that constructed in 132 A.D by **Zhang Heng** (Chinese: AD 79-139). This device consisted of a large brass urn, firmly anchored to the ground which contained a torsion balance, somewhat like a child's see-saw, which would move at the slightest vibration. When vibrated, the balance would then tilt in the appropriate direction and push a small sphere out of the mouth of a dragon figure sculptured on the side of the urn. The sphere would then drop into the mouth of a frog figure below with each frog figure placed at the main points of the Chinese compass. This would indicate the direction of the earthquake.

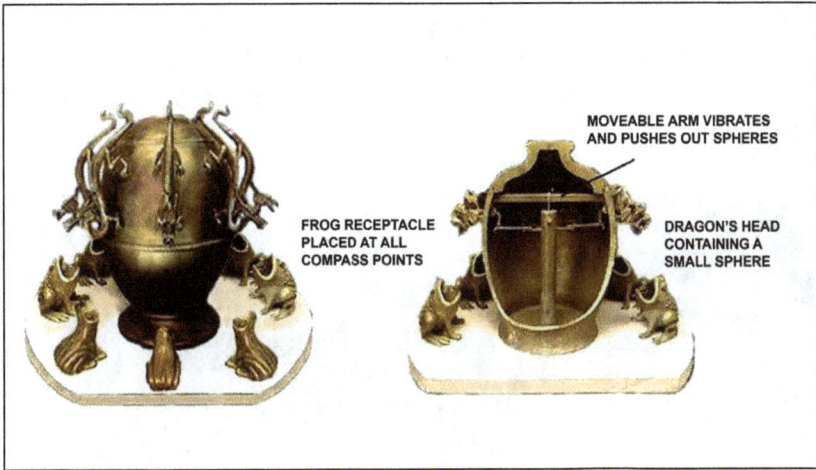

FROG RECEPTACLE
PLACED AT ALL
COMPASS POINTS

MOVEABLE ARM VIBRATES
AND PUSHES OUT SPHERES

DRAGON'S HEAD
CONTAINING A
SMALL SPHERE

Figure 3.7: Diagram showing the early seismoscope of Zhang Heng

In early modern seismometers, a large mass was suspended or hinged with some form of scribe attached to it, such as a pen, ink marker or light beam. This scribe was able to mark a revolving drum called a **kymograph** covered in soot, carbon dust, or wrapped with paper on which the scribe would scrape or draw the trace of the vibrations. This rotating drum and the framework of the seismometer, was anchored into to the bedrock of the seismic station with only the suspended or hinged weight being free to move. When the earth vibrated during an earthquake, the whole frame and its drum would also vibrate, but the pen and suspended mass would remain relatively stationary due to its inertia. Thus, a trace or **seismogram** was produced on the drum as it rotated at a known velocity and time. In modern devices, a sensor is attached to the seismograph which electronically shows the relative movement between the hinged or suspended mass and the bedrock. Any slight motion of the ground will produce an electromagnetic signal which can then be amplified and transmitted by landline, microwave or satellite signal to the monitor or recorder at the seismic station.

Figure 3.8: An early kymograph and attached clock for recording earthquakes

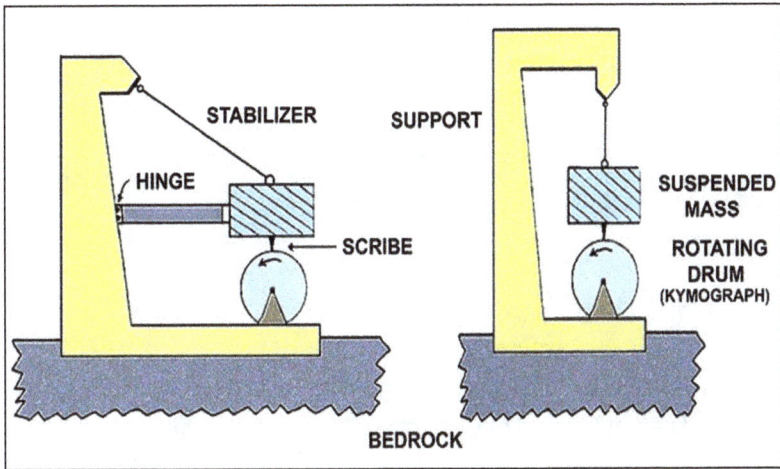

Figure 3.9: Diagram showing the structure of a torsion hinged seismograph (left) and a suspended seismograph (right) to detect different planes of vibration using the principle of inertia

Small seismometers can be taken into the field and placed into protective housing built into the bedrock. These can be connected to a computer or a transmitter which can send any signals back to the receiving station.

Figure 3.10: A portable seismometer about 30 cm. across

Seismograms show the arrival of the several types of earthquake waves at the receiving station. These can be used to measure the earthquake's energy intensity and also the earthquake's location using triangulation with data from at least two other seismic stations.

Online Video 3.1: An example if an incoming seismogram
Go to https://youtu.be/wp3H6GZ_TdE

There are three main types of seismic wave:

- P-Waves (Primary) are **longitudinal waves** which travel in the same axis as the force which produced them and travel through liquid and solid bodies such as the Earth's mantle and core, arriving at the seismic station before any other wave.

- S-Waves (Secondary) are **transverse waves** which travel at right-angles to the force which produced them and are slower than P waves. They are also **body waves** but they will <u>not</u> pass through liquids as they need solid media for their transmission.

- L-Waves (Love - named after the English mathematician, **Augustus Love**: 1863-1940) are a complex form of transverse wave. They travel on the surface as do **Rayleigh waves** which were predicted by and named after **Lord Rayleigh**, English: 1842-1919, which are both transverse and longitudinal waves. Love waves are slower than P and S-waves but faster than Rayleigh Waves. Together, the two surface waves create a rolling motion around the Earth's surface, arriving last of all and causing much damage.

All three waves start out at the focus together, but travel at different speeds and may be reflected, refracted or bent, filtered as S-waves will not pass through liquids, and may change into other types of waveform as they pass through the Earth and meet boundaries of different rock types. In general, earthquake waves travel faster at depth where the material is denser.

Changes in animal behaviour for immediate warning of an impending earthquake at a local level has also been reported, especially in China and Japan where large earthquakes are relatively common. Whilst unusual behaviour in domestic and wild animals such as dogs, horses, cows, chickens, primates, snakes and even ants have been recorded, further studies into this form of early warning are needed.

3.5 Finding the Source of Earthquakes

Knowing the velocity of the various earthquake waves from previous observations and measurements, and the times at which they arrived at the seismic station, the distance from the earthquake's focus to a seismic receiving station can be calculated.
Consider the following seismogram:

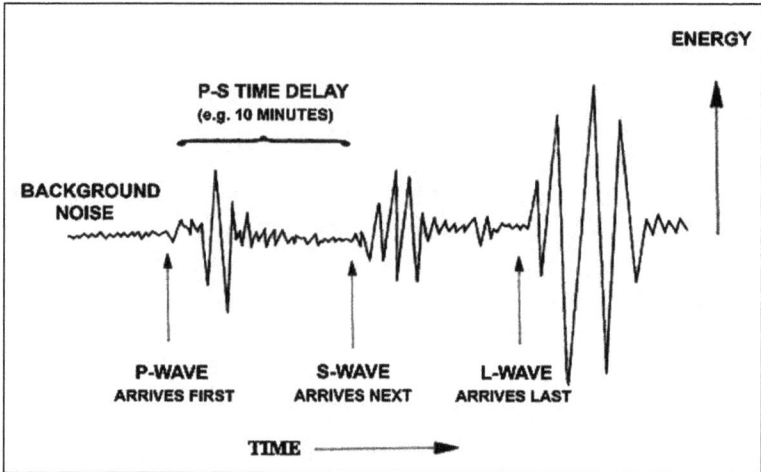

Figure 3.11: a typical seismogram

This typically shows the three main seismic waves which have arrived at the seismic station. All three waves would have started together from the source of the earthquake at its focus. The P and the S waves have travelled through the Earth and the L wave has travelled around its surface within the crust. Because the P and S waves travel at different velocities, it is possible to use the difference in their arrival time as a measure of the distance to the focus. The further away the focus, the longer will be the delay in the arrival of the P and S waves. This is very much like two runners of differing abilities starting together on a long race track. As the runners get away from the starting point, the faster runner will become separated from the slower runner by a greater distance as the race progresses.

In time and with a considerable amount of data about the locations of earthquakes and their distances to seismic stations, seismologists have been able to draw up graphs of **travel time curves** which show the distances from earthquake foci against the time for the earthquake to travel for each type of wave.

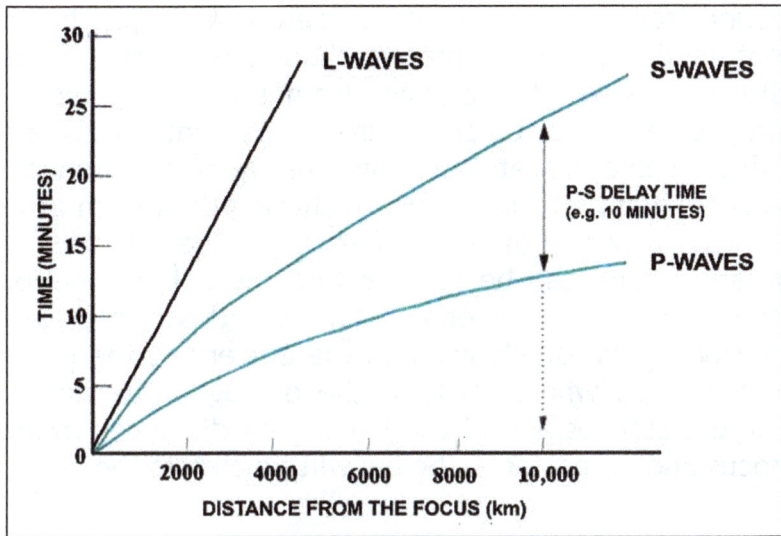

Figure 3.12: Travel Time curves for seismic waves

Knowing the difference in arrival time for the P and S waves, which can be measured directly off any seismogram, the distance to the focus from the receiving station, can be calculated.

For example, consider a difference of arrival time for the P and S waves at a receiving station being 10 minutes.

With this time delay, a travel time curve would show that the distance of the seismograph from the focus would be about 10,000 kilometres away for that receiving station.

Once the direction and distance to the focus has been found from one station, it can be compared to the value obtained by the same method from several other stations. Now, the readings from the seismometer will only give the distance to the focus, not its direction, which could be at any point on a circle around the receiving station. However, in sharing data from at least three other, distant stations, the method of **triangulation** can be used to find the actual location of the focus. In communicating this information, seismologists usually refer to the epicenter, the place on the surface where most of the damage will occur. For large distances across the Earth, the distances from the focus and its epicentre below will be almost the same.

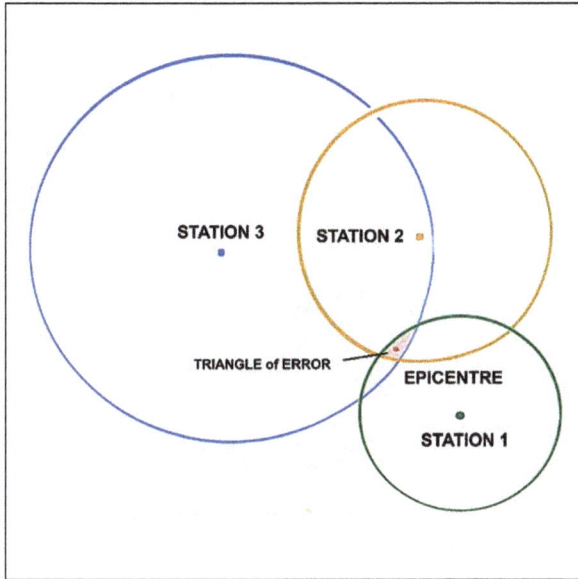

Figure 3.13: Diagram showing triangulation used in locating an epicentre

Remember that the further that the seismograph is away from the epicentre, the greater will be the distance between the arrival of the P-wave and the S-wave because of their difference in velocities. The pathway and measurement of these velocities however, can be very complex because both waves will travel at greater velocities through denser rock.

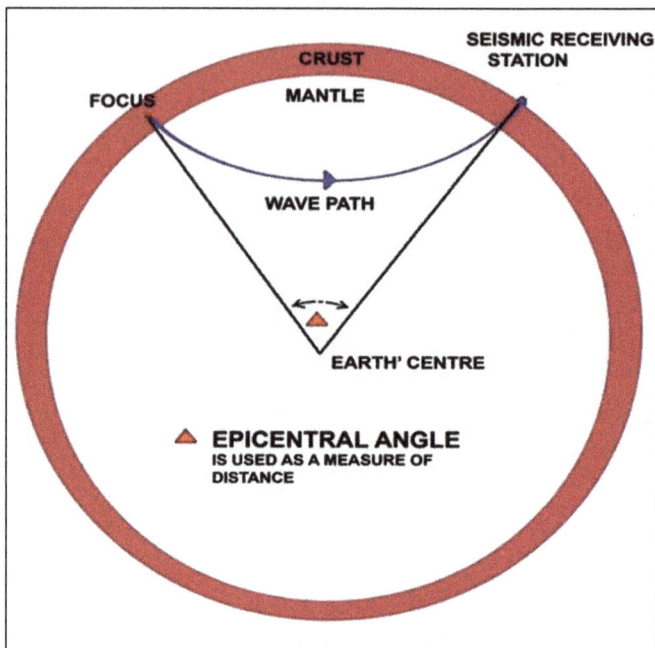

Figure 3.14: Diagram showing that wave pathways are curved due to increasing density of rocks with depth.

3.6 Depths of Earthquakes

Until 1922, all earthquakes were considered to be of shallow origin, but in that year, **Herbert Hall Turner** (English: 1861-1930) of Oxford University, discovered some slight differences between the P waves of seismograms which suggested that many earthquakes had deep sources. This was not confirmed until 1931 when it was noted that some very large earthquakes had surface waves (L waves) with uncharacteristic small amplitudes. In addition, their body waves (P and S waves) were relatively simple without the complications which occurred due to depth by changes at boundaries and increasing density.

Very sensitive seismographs can show these depth changes, indicating that some P waves were reflected off the surface of the Earth very near the epicentre and that these were then transmitted downwards, giving a new version of the P wave termed the pP wave. This pP wave naturally lagged behind the original P wave as it had to travel to the surface and then be reflected. Over a longer distance, the P wave minus the pP wave travel time difference was therefore longer and changed slowly with distance, but rapidly with depth. Using the known distance between the seismograph and the epicentre, and the time delay between the P and pP waves, the depth of the focus could be calculated.

Some calculations for the depth of the foci of earthquakes near deep oceanic trenches gave an interesting pattern, first detected independently by **Hugo Benioff** (American: 1899-1968) and **Kiyoo Wadati** (Japanese: 1902-1995) in 1949. The location of these earthquakes defined the lower part of the oceanic plate as it was pushed down into the Earth. This location was called the **Benioff-Wadati zone**, and it gave some support for the concept of sinking tectonic plates or **subduction zones** in the theory of **plate tectonics**.

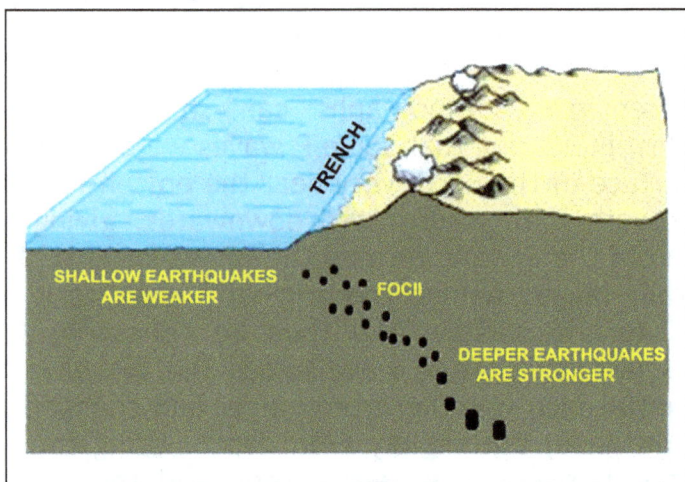

Figure 3.15: Diagram showing a Benioff-Wadati zone at a continental trench

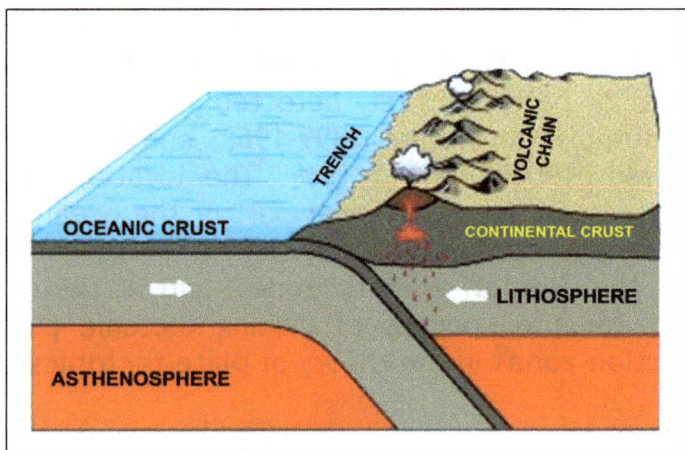

Figure 3.16: Diagram showing the modern interpretation of a Benioff-Wadati zone by plate tectonics

3.7 Measuring the Strength of Earthquakes

The measurement of the amount of energy released by an earthquake can be very complex because of the:

- complex nature of the waves emitted and changes which occur

- variation in rock density through which the waves travel

- sudden boundary changes with refraction and reflection of waves

- nature of the instrument recording the incoming waves

The American seismologist, **Charles F. Richter** (1900-1985), developed a method in 1935 at the California Institute of Technology using the measurement of the amplitude, or the maximum height of the seismic wave from its central axis as an approximation of the earthquake's strength or **magnitude**. His method, which was originally intended for shallow, local earthquakes in California, gave the magnitude of the earthquake (**M**) as a function of its **amplitude** (**A** measured in microns). In simple terms, this can be expressed as:

$$M = \log_{10} A$$

Richter used a standard Wood-Anderson torsion seismometer, and scaled any incoming values as though they had an epicentre 100 kilometres away. For comparison, any other seismologist using this **Richter Magnitude Scale** (M_L) had to use the same type of seismometer. They would also have to calibrate the distance, using the epicentral angle, to 100 kilometres distant. Later, this method was applied to earthquakes on a worldwide scale using different seismometers calibrated to fit the Richter scale.

Values on the Richter scale are given as whole numbers and decimal points, with the difference between the whole numbers being a tenfold magnification, that is, a **logarithmic** function using the powers of ten. Thus a magnitude value of 4.5 has an amplitude ten times greater than a magnitude 3.5 earthquake. An earthquake of magnitude 5.5 would be 100 times that of the 3.5 earthquake and so on. However, this is only a measure of the maximum amplitude of the earthquake waves and only an approximation of the destructive power of the earthquake. This destructive energy could be more like 32 times the difference in energy released between whole numbers because the actual formulas used for the conversion between amplitude magnitude and actual energy release take in several other factors. For example, for body waves (P and S waves), a more accurate relationship is:

$$M = \log_{\text{Base 10}} (A/T) + Q (D, h$$

Where **T** is the period or the time for one vibration of the waves;
Q (D, h) is a correction factor allowing for distance **D** (in epicentral angle degrees); and
h is the depth of the focus in kilometres.

For surface waves (L waves) where the rock layers, water surfaces and rolling nature of the wave gave a more complex situation, seismologists who wish a more accurate measurement of the actual energy of an earthquake, now use the **Moment Magnitude Scale (MMS)** introduced by Caltech in 1979. This new scale was developed largely through the work of **Beno Gutenberg** (German-American: 1889 -1960) to overcome problems which the Richter scale had with very long distance earthquake waves and a lack of sensitivity in measuring surface waves. Even though the formulas are different, the new scale still retains a similar logarithmic scale to that of Richter's scale. It is based on the seismic moment, or the energy released when rocks move during the earthquake, which is equal to the rigidity of the Earth multiplied by the average amount of slip on the fault and the size of the area that slipped.

3.8 The Mercalli Scale of Intensity

Another way of describing seismic events is by using the **Modified Mercalli Scale** of earthquake intensity, or the effects of the degree of surface vibration. Named after the seismologist, **Giuseppe Mercalli** (Italian: 1850-1914), this is a useful scale, as it uses a series of vivid descriptions for the effect of surface shaking on the surroundings. This gives a better but non-scientific understanding of the total effect of the earthquake on places of habitation and enables comparisons from one place to another.

MERCALLI NUMBER	NAME	DESCRIPTION
I	INSTRUMENTAL	detected only by seismographs
II	FEEBLE	felt by a few people
III	SLIGHT	felt by people in buildings
IV	MODERATE	loose objects moved
V	RATHER STRONG	dishes break, objects fall over
VI	STRONG	plaster cracks, furniture moves
VII	VERY STRONG	walls crack
VIII	DESTRUCTIVE	chimneys fall, old buildings collapse
IX	RUINOUS	houses damaged, ground cracked
X	DISASTROUS	most buildings destroyed, landslides
XI	VERY DISASTROUS	large cracks in the ground, landslides
XII	CATASTROPHIC	total destruction, ground heaves

Table 3.3: The Mercalli scale of earthquake intensity

From such descriptions, seismic maps can be drawn up which show places of equal earthquake intensity called **isoseismals**:

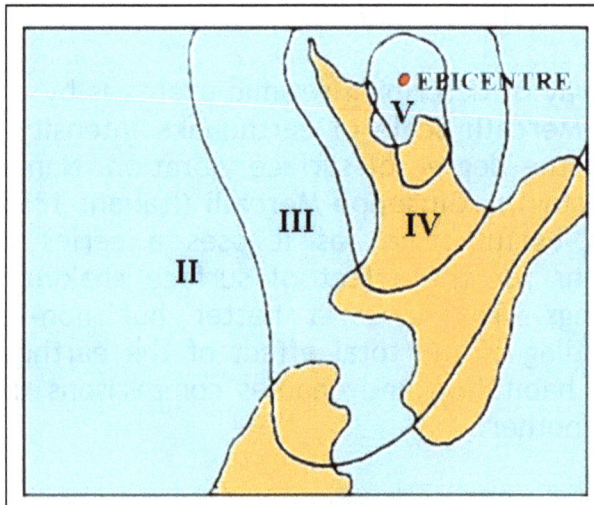

Figure 3.17: An isoseismal map

Whilst it only an approximation, some comparison for popular communication purposes can be loosely made between the Richter scale and the modified Mercalli scale:

RICHTER MAGNITUDE	MERCALLI INTENSITY
1.0 - 3.0	I
3.0 - 3.9	II - III
4.0 - 4.9	IV - V
5.0 - 5.9	VI - VII
6.0 - 6.9	VII - IX
7.0 and higher	VIII or higher

Table 3.4: Richter and Mercalli scales compared

Unfortunately, the modified Mercalli scale, as a non-scientific and subjective set of descriptions is not favoured in popular communications by the Media. The Richter scale is often quoted by the press to give some concept of the size of an earthquake and therefore the damage and loss of life. A glance at the previous table of recent major earthquakes (Table 3.2) will show that many earthquakes which have caused major damage and loss of life were often lower in value of those which had fewer fatalities. For example, the November 26th, 2003 earthquake in southeastern Iran had a magnitude of 6.6 and killed 31,000 people, but that of 11th April, 2012 in northern Sumatra, Indonesia with a magnitude of 8.6 had no fatalities even though both areas have large populations. This is because there are many other factors to consider, such as the concentration of people at the location, the terrain, the state of buildings and time of day. Iran is mountainous and would have many landslides and most of their buildings would be of multiple stories and made of mud brick which would quickly collapse and cause death. In northern Sumatra, Indonesia, the land is less steep and most of the houses would be single story and built of timber this would reduce the number of deaths.

3.9 Earthquake Disaster Management

Earthquakes are even less predictable than volcanic eruptions and their effects can be more disastrous due to:

- falling buildings which cause the majority of deaths

- shattering windows which explode glass out of their frames and drop broken panes down from above

- damaged infrastructure such as roads, bridges, power lines and water supplies

- fires due to broken fuel (especially gas) lines

- landslides which can cover towns, crops and roads

- tsunamis which are the major hazard on low-lying coastal areas

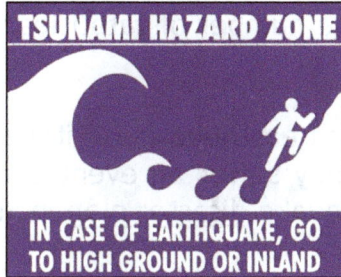

TSUNAMI HAZARD ZONE

IN CASE OF EARTHQUAKE, GO TO HIGH GROUND OR INLAND

Figure 3.18: Tsunami warning sign, on a beach at California, USA

During an earthquake there is little time to consider anything other than to escape the immediate hazard of collapsing buildings or threat of a tsunami. If one is inside a building, it is wise to stay inside and shelter under an archway, under a solid table or bench or in a corner where there will be less chance of being struck by falling masonry. Once the initial shaking has stopped then it may be possible to evacuate to an open space well away from buildings, bridges, overpasses and power lines. If on the coast, then one must get to high ground in case of a tsunami.

The hazards associated with earthquakes often do not stop once the first shock has stopped. There may be aftershocks which can be more dangerous than the initial shock, as the pressured rocks re-adjust and release pressure. Moreover, the entire surrounding infrastructure may be damaged causing a complete lack of shelter, water, food and communication. There are further health issues as well, especially if the earthquake has occurred in remote mountainous areas where the climate can be severe. Death by exposure, lack of water and food and disease from contaminated water supplies usually add to the number of fatalities of the earthquake event.

Preparation for an earthquake is vital for those who live in areas which may suffer an event one day. This would include an earthquake disaster plan including:

- What to do immediately, such as knowing the location of the strongest part of the building (usually an arch, a very small room or a corner of a large room) and how to get there quickly. This would be the best place to find shelter. In many school and office situations, the most immediate place of shelter would be under a desk.

- Where to go after the first shock is over. This would be a known evacuation exit plan and shelter, or open space well away from the building. An evacuation route should be well-known or clearly displayed for any guests or clients as well as family.

- Where to go if a tsunami may occur. This would be the next step after surviving the initial shock if one is in a coastal region and many earthquake-prone coastal regions have tsunami alerts. The best place would be the nearest, open high ground or if that is too distant, a high, solid structure which has survived the initial shock intact.

- What to do next is always an important question after one has escaped the initial shock and/or tsunami. Locating relatives and friends may not be easy, as communication systems such as mobile phones would not be operating, and there would be great disruption to movement and little if any, transportation available. Furthermore, there may not be food, water and shelter available and emergency aid may not arrive until several days. The best plan is to stay in the agreed evacuation site unless directed to move to other places of shelter by the authorities.

- Having an emergency kit which is stored in a structurally-strong place, such as the entrance archway or outside shed would be useful, ready to be grabbed on the way to the evacuation site after the initial shock is over. This kit may contain:

    - non - perishable food, cooking and eating utensils, including can-opener

    - cooking fuel in small amounts for a few days and sealed in appropriate containers. This is also a fire hazard. The best cooking fuel consists of hexamine pellets sold as "Fire Starters" which can be stored in a sealed plastic container

    - water for at least two days and a water purification kit as local water supplies may be contaminated with sewerage or animal bodies which can lead to health problems if used for drinking

- shelter, such as a small tent or ground sheet. In cold areas this could be an aluminised thermal blanket sold in many hiking stores

- first - aid kit and prescription medicines

- plastic bags for water storage and waste

- personal hygiene items

- warm and waterproof clothes and sleeping bags

- portable radio, headlamp, and extra batteries

- pocket knife, whistle, matches (or cigarette lighter), duct tape, small signal mirror and gloves.

All of these compact items can be easily stored in a small, waterproof pack which can be quickly snatched on the way to the evacuation area. It can be packed with enough supplies for a small family for two days.

Figure 3.19: Earthquake secure zone in an archway of a posada (inn), Arequipa, Peru (it reads: "secure zone in case of earthquakes")

Chapter 4: The Interior of the Earth

4.1 Introduction

Scientists have only penetrated a small distance into the surface of his planet despite great advances into space. The deepest drill hole, on the Kola Peninsula of Russia, is 12,262 metres deep. The deepest mine in South Africa is 3900 metres deep and the deepest natural cave is in Abkhazia, and is only 2197 metres deep.

From the chaos of early superstitions about the interior of the globe, a few scientific ideas gradually emerged by the 19th century:

- From Newton's **Law of Universal Gravitation**, the mass and, therefore, the density of the Earth could be calculated by astronomical data. This Law states any mass attracts every other mass in the universe using a force that is directly proportional to the product of their masses and inversely proportional to the square of the distance between them.

- The average density of the Earth is about 5.5 g/cc and with a comparison to that of surface rocks of only about 2.5 g/cc, led to the belief that the interior must contain very dense material to give such a high average value.

- The Earth's magnetic field, used by mariners for centuries in navigation, suggested that there must be considerable iron within the Earth.

- Heat measurements in deep mines indicated that heat increased with depth (about 30°C/km) such that the interior of the Earth should contain very high temperatures sufficient to melt rock. This was supported by volcanic observations.

- Pressure and density generally increase with depth and this would explain some of the other findings about the Earth's mass.

4.2 A Layered Earth

Most of these theories, however, still did not give any structural detail, but only generalized physical data about the Earth as a whole. It was not until the late 19th century and early 20th century that evidence for a layered earth began to emerge from the study of earthquakes.

Early in the 20th century, the Croatian seismologist, **Andrija Mohorovicic** (pronounced *Mo – horo- veech – ich*: Croatian: 1857-1936) of the University of Zagreb, noticed that there were two sets of similar P and S waves arriving at his seismic station for the same earthquake. This had occurred on the 8th October, 1909 with a stronger set of waves arriving at recording stations after a weaker set of identical waves for the same earthquake. He suggested that the second set of stronger waves had travelled directly from the focus to the stations as expected but that the first, weaker set had been refracted through a denser layer below the surface. As the velocity of P and S waves increases with density, the weaker waves had travelled in the denser medium, caught up and passed the usual, stronger set of waves which arrived slightly later. It was also noted that this effect was more pronounced when the earthquake foci were at greater distances to the receiving station.

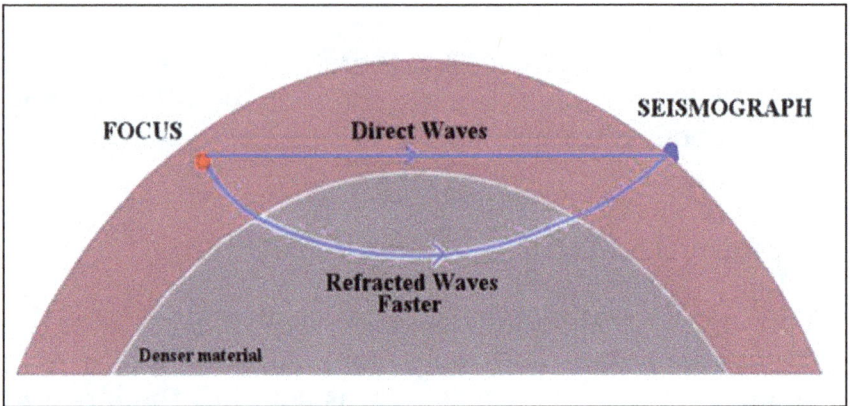

Figure 4.1: Diagram showing how a second wave could reach a seismograph faster

This boundary at which the waves were refracted became known as the **Mohorovicic Discontinuity,** or Moho, after the famous seismologist, and it varies in depth below the surface from about 12 kilometres below oceans, to about 70 kilometres below continents. The concept of a non-layered Earth now changed to one which had a top layer, or **crust** made up of less dense rocks such as lighter volcanics, sedimentary rocks and metamorphic rocks. Below this is the **mantle** consisting of denser rocks rich in ferromagnesian minerals, especially **peridotite**. The crust and the upper mantle together form the solid **lithosphere** which is broken up into several plates covering the Earth's surface.

Using more refined equipment and new knowledge of the pathways of earthquake waves through the Earth, another boundary within the crust of continents was soon discovered. This was located between rocks of lighter density, called the **SIAL** (for **Si**lica and **Al**uminium) in the upper crust, and those below of greater density, called the **SIMA** (for **Si**lica and **Ma**gnesium). This is called the **Conrad Discontinuity** (after **Víctor Conrad**, Austrian: 1876-1962) and is the indistinct boundary between the lighter, granitic rocks of the continents and the denser, mafic rocks, such as basalt, of the lower crust below. In recent years however, some geologist have suggested that this may simply be a change between an amphibolite facies to granulite facies of regional metamorphism which may exist within the deeper crust. It is not found below oceanic regions.

In the 1920's, Beno Gutenberg discovered that, at certain depths, which varied from about 100 kilometres to about 250 kilometres, the velocities of earthquake waves suddenly slowed down. This place was called the **low velocity zone,** and it forms the base of the **aesthenosphere,** that part of the upper mantle which is solid but has places, such as below mid-ocean ridges which may be liquid. The low velocity zone defines the bottom of the tectonic plates and is probably a semi-liquid zone upon which these plates move.

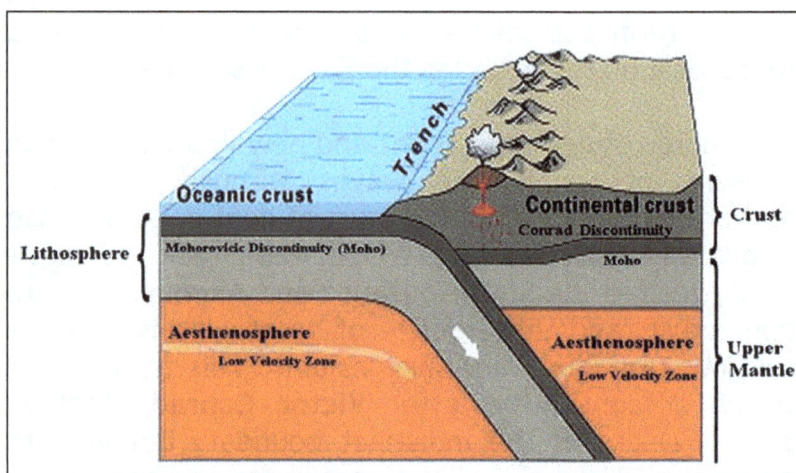

Figure 4.2: Diagram showing the upper layers of the Earth

At about 2898 km below the surface, earthquake waves suddenly suffer another sudden drop in velocity. In addition, S waves suddenly disappear whilst P waves continue but are refracted. This marks another major discontinuity, the **Gutenberg Discontinuity**, which marks the boundary between the lower mantle and the outer core.

The reflection of P and S waves off the outer core at certain critical angles, and the refraction of P waves through the outer core, causes places on the Earth's surface which do not receive any waves from some earthquakes. The only indication in these places that there has been an earthquake on the other side of the Earth, are the surface waves (L waves) which travel around the surface. The regions where there are no P or S waves are called **shadow zone**s and they lie between 103° and 143° of epicentral angle from the epicentre of the earthquake (i.e. 103° and 143° between the earth's centre and the epicentre).

The fact that S waves require a solid medium for transmission, and that they do not travel through the outer core, suggests that this section may be a liquid. The velocities and refractions observed throughout the mantle also suggest that the outer core contains mainly rocky material of high density.

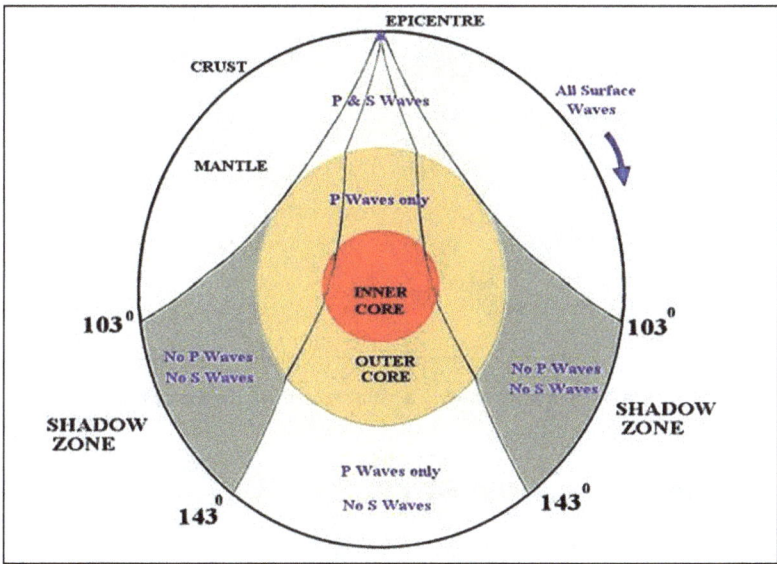

Figure 4.3: A very simplified view of earthquake waves travelling through the Earth to give shadow zones. Within the cores there are many wave transformations, refractions and reflections.

Slight traces of P and S waves within the inner core suggest that it may be solid. Studies of **meteorites**, which are either rocky or metallic nickel-iron, suggest that the inner core may also be composed of nickel-iron. This evidence is supported by the Earth's having a magnetic field and that and density calculations for the whole of Earth's volume agree with its inner core being of nickel-iron.

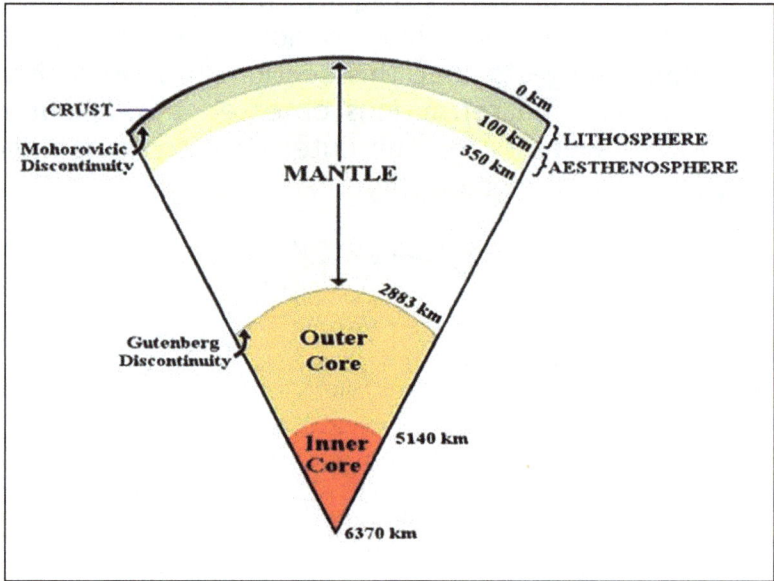

Figure 4.4: Diagram showing a model for the interior of the Earth

4.3 The Shape of the Earth

Geodesy is the study of the shape of the Earth. **Aristarchus of Samos** (Greek: 310-250 BC) noted the curvature of the shadow of the Earth as it moved across the moon during a lunar eclipse, and suggested that the Earth was also a sphere. Later, **Eratosthenes of Cyrene** (276 – 194 BC) in Egypt, measured the radius of the Earth assuming that it was a sphere. He had made the observation that at noon during the summer solstice, the sun was directly overhead at Syene (or Cyrene, modern Aswan) but at Alexandria further north, at exactly the

120

same time, the sun gave a shadow of a little over seven degrees to the vertical. By knowing the exact distance between Syene and Alexandria, he was able to use simple trigonometry to calculate that this distance represented one fiftieth of the Earth's circumference.

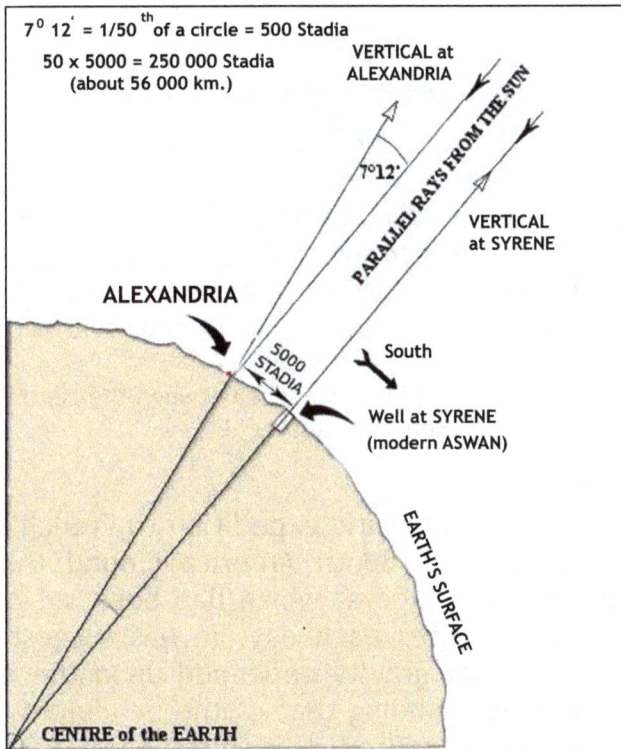

Figure 4.5: Eratosthenes' method of finding the Earth's circumference.

By the Middle Ages, similar geometrical methods, using stars as reference points, led Greek, Arab and Dutch astronomers to the same conclusion - that the Earth was a sphere.

Around 1665, **Sir Isaac Newton** (English: 1642-1727) proposed his theory of **Universal Gravitation** which stated that all masses are attracted to each other depending upon their mass and the inverse of the square of the distance between them:

$$ F \; \alpha \; \frac{M_1 \, M_2}{d^2} $$

where α means "is proportional to"
F = force due to gravity;
G = Universal Gravitation Constant;
M₁ and **M₂** are masses (one of which could be the mass of the Earth); and
d² = square of the distance between the centres of the two masses.

In 1672, a French scientific expedition to French Guiana, near the Equator in South America, found that their accurate pendulum clocks, which had been set at Paris, lost about 25 minutes each day. It was suggested that Newton's idea that gravity depended upon the distance between masses including that of the pendulum and the Earth, may have caused this discrepancy. This was because the distance to the centre of the Earth at Paris was different to that at the Equator. From his calculations, Newton proposed that the Earth's radius measured at the Equator should be one part in 230 longer than the radius measured at the poles. The concept that the Earth was not perfectly spherical, as suggested by the Ancient Greeks, but was more of an elongated sphere became a hypothesis worth testing by many astronomers. In the 18th century, **Giovanni Cassini** (Italian-French: 1625-1712) provided data to suggest that the Earth was indeed elongated along the equatorial axis. To solve this riddle, elaborate French expeditions were sent to Lapland and to Peru in 1730. After many years of hard exploration, observation and calculation, inconclusive values of between 1 part in 178 and 1 part in 266 were reported for an Earth radius elongated at the Equator. Closer to modern times, astronomical observations made in 1948 give a value of 1 part in 297 for an equatorial elongated Earth.

Today, the ellipticity and other effects of the Earth's gravitational pull on the orbits of artificial satellites, the use of global positioning satellite triangulation, and very accurate Earth-based **gravimeters**, which measure the gravitational pull, give a very accurate idea of the size and shape of planet Earth. Data from these observations indicate an ellipticity of the surface of the Earth of about 0.00335 (i.e. a ratio of 1: 298).

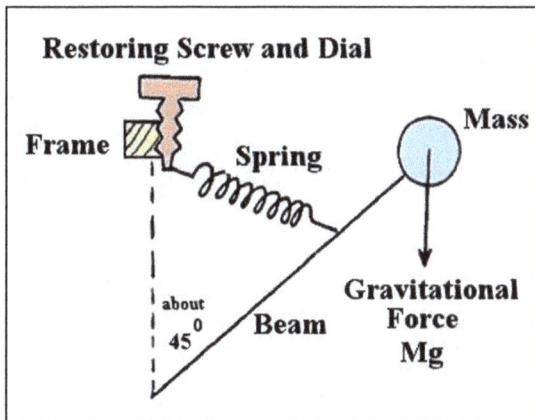

Figure 4.6: Diagram showing the basic principle of a gravimeter

The final conclusion from past and present data is that the Earth is an **oblate spheroid,** being generally flattened at both the North and South Poles, with a bulge at the Equator. The Earth is rather pear-shaped, with more flattening at the South Pole than at the North. As well as its general ellipticity there are several minor variations, such as hollows near India, South America and the South Pole and bumps in Western Europe and near Papua New Guinea. It is thought that equatorial bulge and the flattened poles can be explained by the greater amount of **centrifugal force** at the Equator due to the rotation of the Earth.

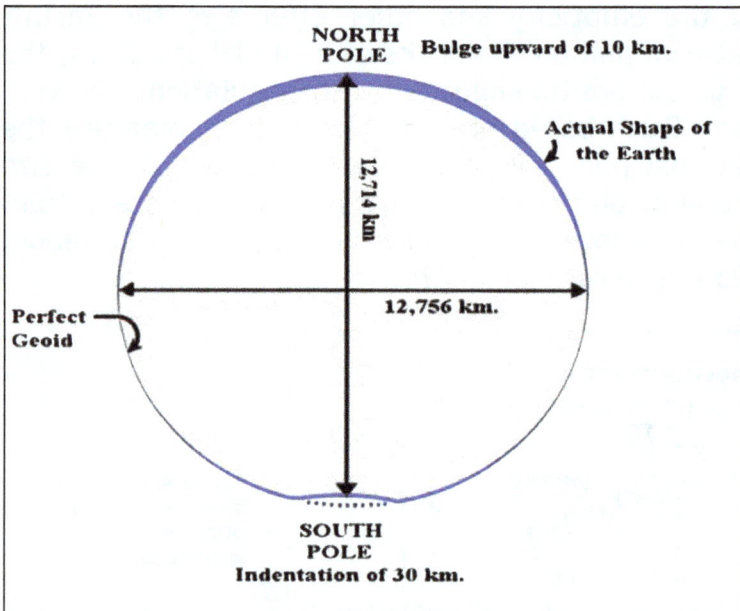

Figure 4.7: Diagram showing the shape and size of the Earth

Summary

1. The Earth's surface is no longer considered as a solid mass but maintains a balance between vertical movements such as mountain-building or orogeny, and the sinking of sedimentary basins. This general maintenance of a vertical equilibrium is called isostasy.

2. Evidence for the gradual uplift/sinking of the surface comes from the formation of marine rock platforms and satellite GPS measurements, whereas more dramatic evidence for the dynamic nature of the Earth comes from earthquakes.

3. Applied pressure to solid rock or stress, will cause deformation and volume change or strain, until it reaches its elastic limit and the rock will break.

4. Hooke's law for solids states that within the elastic limit, stress is proportional to strain. That is, when a rock is pressured it will uniformly deform until it reaches the elastic limit.

5. Different rocks behave in different manners to applied force depending upon their composition and structure – a term called rheidity. Some rocks are ductile and will easily bend e.g. many sedimentary rocks, whereas others are brittle and easily break e.g. crystalline rocks.

6. Joints within rocks often indicate the direction of the forces which caused the rock to crack. Conjugate joints are usually at a sixty-degree angle to each other with the direction being the bisection of this angle whereas

columnar jointing is caused by a shrinking inwards as cooling or drying occurs.

7. When there is movement at a fault, a near-vertical cliff or fault scarp is often formed. Bare rock at the face of this scarp may show scratches called slickensides or within the fault line there may be ground rock material or fault breccia.

8. Block mountains or horsts, and sunken valleys with steep, parallel sides or grabens, are caused by faulting usually due to being pulling apart or tension. Fold mountains are formed by compressional forces.

9. When compressed below their elastic limit, large areas of rock can be folded into simple and complex shapes from microscopic size to large fold mountain ranges.

10. Pyroclastics are broken material erupted out of volcanoes. This material or tephra, can range in size from microscopic dust to large blocks.

11. Volcanoes can give out large volumes of gases from their fumaroles or small gas vents, including steam,carbon dioxide, sulfur dioxide, nitrogen and other gases.

12. Volcanoes are extrusive igneous structures because they form on the Earth's surface. Volcanoes can be formed from ash as cinder cones, lava as with rhyolite domes and shield volcanoes, or a combination of successive ash and lava as strato-volcanoes or composite volcanoes.

13. Volcanoes may be active and regularly erupt, dormant when they have not erupted for some time, or extinct when they have not erupted in historical times.

14. Active volcanoes occur at the edges of the Earth's plates as one plate sinks below another at a subduction zone, where one plate has split and is now spreading apart at a mid-ocean ridge, or sometimes in the middle of a plate above a deep convection current, hot spot or plume in the mantle.

15. Earthquakes occur at a focus when there is a sudden breakage of rock or movement along a fault plane or plate margin due to the rock being stressed and deformed then suddenly releasing energy. This concept is called the elastic rebound theory.

16. Earthquakes can be slight as small tremors, or major events causing large-scale damage by the collapse of buildings, roads and dams, landslides and ground upheaval. Tsunamis formed by vertical movement under the sea also cause massive damage and loss of life along coastal regions.

17. Earthquakes can be detected by seismographs, also called seismometers, which measure the relative motion between the Earth's surface and a heavy, suspended weight which resist any change in motion due to its inertia. Seismograms are patterns produced from waves reaching a seismometer.

18. There are several types of earthquake waves produced at the focus: P waves are longitudinal which pass through solids and liquids; S waves are transverse waves and do <u>not</u> pass through liquids; and L waves are complex waves and travel around the Earth's surface.

19. The Richter scale is a measurement of the magnitude of an earthquake using a logarithmic scale of powers of 10 for the amplitude of the earthquake such that the whole number difference between values e.g. 4.5 and 5.5. is equal to a ten-fold amount. This is only an approximation of the relative sizes of earthquakes.

20. Seismologists who wish a more accurate measurement of the actual energy of an earthquake, now use the moment magnitude scale (MMS) introduced by Caltech in 1979 to overcome problems which the Richter. Even though the formulas are different, the new scale still retains a similar logarithmic scale to that of Richter's scale and is based on the seismic moment, or the energy released when rocks move during the earthquake, which is equal to the rigidity of the Earth multiplied by the average amount of slip on the fault and the size of the area that slipped.

21. The modified Mercalli scale measures intensity or the amount of shaking at the epicentre, on a twelve-point scale subjectively describing damage.

22. Earthquake waves can be traced right through the earth and by noting where and how they emerge at seismic stations around the world. Since the turn of the 20th century, various boundaries or discontinuities, have been found showing that the Earth has several layers which are the crust, mantle, outer core and inner core.

23. These boundaries include: the Mohorovicic Discontinuity which is the boundary between the lighter, rock of the crust and the more dense rocky mantle below; the low-velocity zone as the boundary between the more solid lithosphere of the upper mantle and the thin region of the upper mantle of ductile material able to flow called the aesthenosphere below; and the Gutenberg Discontinuity between the mantle and the denser outer core below.

23. Geodetic studies of the Earth using astronomy, gravitational calculations and satellite measurements have shown that the Earth is an imperfect oblate spheroid i.e. an elongated sphere, slightly flattened at the poles and extended out at the Equator. The elongation is only very slight, being only 1 part in 298 extended.

Practical Tips

1. Analysing joint patterns requires considerable patience. At least several hundred measurements of the strike of joints on an exposed rock platform are needed before these can be used to construct a rose diagram showing any pattern of joint orientation. Often there are several separate episodes of forces causing separate sets of joints.

2. When in a country prone to earthquakes, make sure that the local safety procedures are known and look for any sign posts that may indicate shelter.

3. Basic earthquake drill includes:

   - drop, cover beneath a strong desk or a small room or support corner and hold on.

   - evacuate only when shaking has stopped.

   - carefully exit the building or structure remembering that many fatalities are caused by falling debris including glass, even after the shock has passed.

   - do not light fires such as cigarette lighters in case there are gas leaks.

   - go to the nearest designated safety zone or to the nearest open space away from buildings.

- safely help others who can easily be reached.

- keep out of the way of emergency workers unless asked to help or if you need help.

- if uninjured try to get home or connect with family safely.

4. Remember that in most severe earthquakes, roads, telephones and cell phones as well as other urban utilities will possibly not be functioning.

5. Tsunamis are caused by vertical motion of the seafloor, so if an earthquake is experienced in coastal areas assume that a tsunami is possible so seek high ground.

6. When in an active volcanic area, take note of local advice and do not venture beyond safe limits. Craters and their edges can be very unstable as they are a thin basaltic lava field. Eruptions can be very sudden and give little warning.

7. The effects of an eruption can be felt at large distances from the vent. Lahars, nuée ardentes and fast lava flows will flow down river valleys faster than vehicles can drive. Seek high ground.

8. Falling ash, even at a great distance can cause suffocation and building collapse.

9. Even after an earthquake or extreme volcanic eruption there will be problems such as lack of water, food and shelter so if living in areas prone to these natural disasters, keep alert to official media warnings and have an evacuation plan and kit in a convenient place.

Multichoice Questions

1. This question refers to the following diagrams of faults:

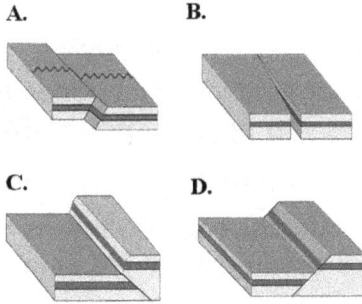

A.   B.

C.   D.

The diagram which best represents a fault caused by compressional forces within the rock is:

    A. A
    B. B
    C. C
    D. D

2. Pyroclastic rocks are igneous rocks formed from:

    A. Lava flows
    B. Lava and ash layers
    C. Broken fragments
    D. Intrusive rock layers

3. Lava which is very fast-flowing probably has:

   A. Low quartz content;
   B. Low mafic content;
   C. High silica content
   D. High feldspar content

4. The type of volcano usually found near plate subduction zones at ocean trenches is:

   A. Basaltic
   B. Rhyolitic
   C. Granitic
   D. Andesitic

5. The following photographs are of different types of volcano. The one which is a shield volcano is:

A.

B.

C.

D.

   A. A
   B. B
   C. C
   D. D

6. The pattern of wavy lines produced on an instrument which measures earthquake waves is called a:

    A. Seismograph
    B. Seismogram
    C. Seismometer
    D. Seismic record

7. The system of communicating the severity of an earthquake by giving descriptions of the damage use the:

    A. Richter scale
    B. Mercater magnitude scale
    C. Modified Mercalli scale
    D. Earthquake scale

Questions 8 and 9 refer to the following graph which shows the travel times for seismic waves and depth below the Earth's surface:

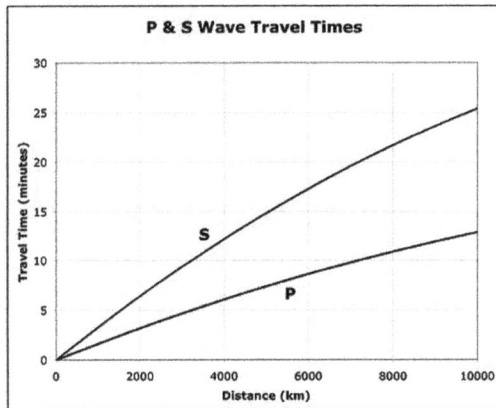

P & S Wave Travel Times

8. At a distance of 6000 km, how long in time will an S wave arrive after the P wave?

> A. 5 minutes
> B. 8 minutes
> C. 12 minutes
> D. 17 minutes

9. At equal depths in the earth, S waves travel:

> A. Always faster than P waves
> B. Always slower that P waves
> C. Always at the same rate as P waves
> D. Sometimes faster and sometimes slower than P waves

10. This question refers to the following cross-section through a part of the Earth.

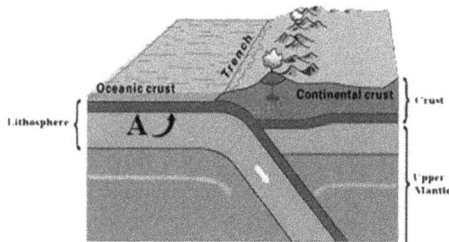

The boundary labelled A. is most likely the:

> A. Mohorovicic Discontinuity
> B. Aesthenospheric Discontinuity
> C. Conrad Discontinuity
> D. Gutenberg Discontinuity

Revision and Discussion Questions

1. The following diagram is a view of a rock platform taken from above. It shows a particular type of joint pattern:

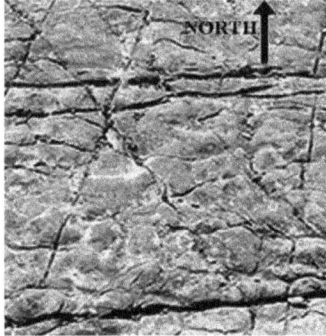

   What name is given to this type of pattern?
   Suggest the directions of the applied force causing the pattern.

2. What is Hooke's Law? How does it relate to the possible structures which may form in rock whe pressure is applied? What factors could also b considered when thinking about which structures coul form in the rock layers other than pressure?

3. What would be some indicators that a local area was once a site of volcanism?

4. Distinguish between P, S and L waves. Which of these waves would <u>not</u> travel through water? What is the significance of that?

5. Discuss the terms magnitude and intensity with reference to the suitability of each system as a means of communicating the relative sizes of earthquakes.

6. Countries like Australia only have relatively mild earthquakes but countries such as Papua New Guinea and New Zealand have major earthquakes. Why?

7. Briefly distinguish between:

   (a) crust and mantle
   (b) seismograms and seismographs
   (c) focus and epicentre
   (d) joints and faults
   (e) throw and heave of faults

8. What is rheidity? How does it relate to the type of structures which may occur in rock which is being compressed?

9. What is a shadow zone? What does it show about the Earth's core?

10. This question refers to the travel-time graph for earthquakes given below:

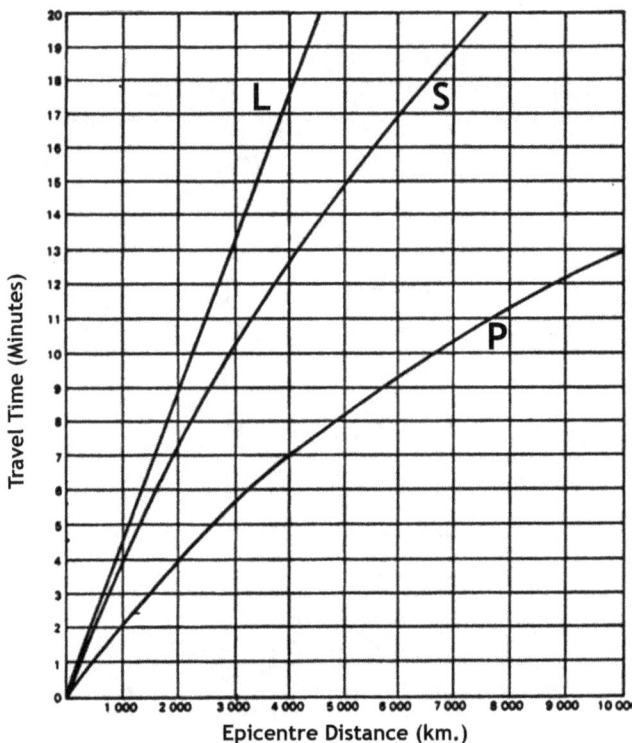

Three seismic stations received seismic wave patterns for the same earthquake:

STATION	P ARRIVAL TIME (HRS. MIN. SEC.)	S ARRIVAL TIME (HRS. MIN. SEC.)
A	11.22.00	11.25.00
B	11.23.00	11.27.00
C	11.26.00	11.30.30

(a) How far is station A from the epicentre?
(b) Give two reasons from the data why station A Is probably closer to the epicentre?

11. Discuss the role of Earth satellites in the investigation
    of the Earth's shape.

12. The table of recent earthquakes given previously in this book shows some very large earthquakes causing very few fatalities and some which are much smaller (remember the logarithmic scale of the Richter numbers) which have caused very large numbers of fatalities. Explain?

13. Use the Internet to make a critical evaluation of the

   (a) earthquake hazard potential,
   b) tsunami hazard potential and the
   (c) volcanic hazards in your location

   Discuss this with your colleagues.

14. Give examples of how research and observation from several scientific disciplines have led to an understanding of the nature of the Earth's interior and the Earth's shape.

15. Use the internet to <u>assess</u> the volcanic hazard potential of each of the following places:

        (a) Mexico City
        (b) Naples, Italy
        (c) Kalapana, Hawaii
        (d) Kagoshima, Japan
        (c) Cairns, Australia

Answers to the Multichoice Questions

**Q1. C Q2. C Q3. A Q4. D Q5. D Q6. B Q7. C Q8. B
Q9. B Q10 A**

Reading List

Australian Geographic. (March, 2011) *The 10 Biggest Earthquakes in History.*

Decker, R. W. & Decker, B. (1991). *Mountains of Fire: The Nature of Volcanoes.* Cambridge University Press. p. 7. ISBN: 0-521-31290-6.

Hamilton, J. (2010). *Volcano.* Compton Verney, UK. ISBN-10: 0955271959

Hull, E. (2015_. *Volcanoes Past and Present.* Colorado Springs., CO.   CreateSpace Independent Publishing Platform. 142 pp. ISBN-10: 1505584779

Kranendonk, V. & Martin, J. (2011). Onset of Plate Tectonics. *Science* *333* *(6041):* *413-414.* *doi:10.1126/science.1208766.* PMID 21778389.

Marti, J.& Ernst, G. (2005). *Volcanoes and the Environment.* Cambridge University Press. ISBN: 0-521-59254-2.

NOAA, 2015: *Interactive Map of Historical Tsunamis from NOAA's National Geophysical Data Center.* http://maps.ngdc.noaa.gov/viewers/hazards/

Richter, C.F. (1958). *Elementary Seismology.* San Francisco: W.H. Freeman.

Scrope, G.P. (2009). *Volcanoes.: The Character of Their Phenomena, Their Share in the Structure and*

*Composition of the Surface of the Globe, and Their Relation to Its Internal Forces.*University of Michigan Library. 522 pp. ASIN: B002KT3FCY

Skinner, B.J., Porter, S.C. & Park, J. (2004). *Dynamic Earth - An Introduction to Physical Geology.* 5th edition". pp 236: John Wiley & Sons, Inc.

Stein, S., Wysession, M. (2009). *An Introduction to Seismology, Earthquakes, and Earth Structure. Chichester: John Wiley & Sons.* ISBN 978-1-4443-1131-0.

Turcotte, D.L. & Schubert, G. (2002). *Plate Tectonics. Geodynamics* (2 edition). Cambridge University Press. pp. 1-21. ISBN 0-521-66186-2.

William H.K. Lee; Paul Jennings; Carl Kisslinger; Hiroo Kanamori (27 September 2002). *International Handbook of Earthquake & Engineering Seismology.* Academic Press. pp. 283-ISBN 978-0-08-048922-3

Key Terms Index
(page numbers in brackets)

acidic magma (39) is molten rock having a higher degree of quartz content.

aesthenosphere (120) from Greek *asthenés* for weak and sphere is the highly viscous, ductile region of the upper mantle of the earth below the lithosphere, at depths between approximately 80 and 200 km below the surface.

amplitude A (105) of a wave is the height of the wave from its central (horizontal level) position.

anticline (33) is an upward fold or section of a fold.

basic magma (39) is molten rock having a composition with very little quartz content.

Benioff - Wadati zone (103) is a planar zone of seismicity of earthquake foci corresponding with the submerging slab in a subduction zone.

body waves (96) are those earthquake waves (such as the p and S waves and their various conversions) which travel through the body of the earth.

brittle (11) are rocks which easily break when pressure is applied and form joints and faults.

caldera (55) is a Spanish word derived from the Latin *caldaria*, meaning cooking pot, is the very large circular depression with sharp edges formed when the top of a volcano collapses when the magma chamber below becomes depleted.

centrifugal force (127) is a reaction force on a spinning or rotating mass which is directed outwards.

Conrad Discontinuity (120) within the earth's crust is the indistinct boundary between the lighter, granitic rocks of the continents (SIAL) and the denser, mafic rocks, such as basalt, of the lower crust below (SIMA).

crust (119) is the upper layer of the earth and consists of lighter rock material of as a great variety of igneous, metamorphic, and sedimentary rocks.

cryovolcanism (73) is a form of tectonism where cold material is ejected in the form of gases and cold liquids such as from the surfaces of moons of some of the outer planets.

debris flow (28) is a rapid movement of regolith and (usually) water downhill.

discontinuity (119) is a boundary between two different layers of the earth's interior.

dormant (38) volcanoes are those which have not erupted for some time but may erupt at any moment. A few volcanoes are constantly active but most have dormant phases.

ductile (11) is the ability of a material to bend and produce folds.

elastic limit (12) is the range of applied force in which a deformed rock will return to its usual shape after the force has been removed.

elastic rebound theory (83) explains how earthquakes are formed by the build-up of pressure with a rock layer which is suddenly released when the layer breaks or moves.

electrical fireballs (90) may sometimes be seen just above the ground as the forces within the ground may disrupt the ionic charges within minerals of the soil and rock causing a concentration of electrical charge.

epeirogeny (2) from the greek e*peiros* – land, and *genesis* for creation, is the general uplift of landmasses.

epicentre (84) is the place on the earth's surface immediately above the source of the earthquake (focus) where the maximum effects are felt.

extinct(38) generally referring to non-existance or no event occuring within living history.

faults (18) are joints along which there has been vertical or horizontal movement or both. This movement can be normal (downward slip due to tension), reverse (upward by compression), strike-slip (sideways relative motion) or hinge (the fault opens laterally).

focus (83) is the place below the earth's surface where an earthquake originates.

folds (29) consists of bended and contorted rock layers due to relatively slow compression.

foot wall (20) is the section or block that is below the fault. it is the solid footing which does not move.

fractures (14) is a term used synonymously with joints to mean a crack along which no movement has occurred.

fumaroles (60) are gas vents which emit steam and other gases in and around volcanic areas.

geodesy (123) is the study of the shape of the earth.

geysers (60) name adopted from the Icelandic original to all hot water and steam eruptions which occur when groundwater seeps down fissures in the hot rock of volcanic areas and then boils explosively.

graben (26) are valleys with steep, parallel sides caused by the downward movement of normal faults on each side.

gravimeters (126) are instruments which measure gravitational pull.

Gutenberg Discontinuity (121) marks the boundary between the lower mantle and the outer core inside the earth.

hanging wall (20) is the section or block above the fault and is the part which moves.

heave (21) is the horizontal displacement of the fault as seen by any a linear feature on both sides of the fault (a marker bed).

Hooke's Law (12) states that within the elastic limit, stress is proportional to strain.

horsts (26) are blocks left behind or pushed up by fault movement.

hypocentre (83) is another name for the focus or origin of an earthquake.

inertia (92) is the property of all masses to resist change in their position or motion.

Intensity (107) is a measurement of the strength of an earthquake using subjective descriptions.

intraplate earthquakes (92) are different in that they occur occasionally within the middle of tectonic plates along old fault lines. They can be very destructive.

isoclinal folds (33) are a series of parallel symmetrical folds.

isoseismals (108) are lines joining places of equal earthquake intensity.

isostasy (3) from the Greek: *isostatic* – same position, is the maintenance of a vertical balance over the earth's surface.

joints (12) are small to very large cracks in brittle rock which have not undergone any movement.

kymograph (93) is a rotating drum with a surface which can be inscribed with traces made by a pen or scraper for recording earthquakes and other changes from the main detector.

lahar (28,56) is a fast-flowing volcanic mud flow which is a mixture of fine ash and meltwater from snow.

LASER geodometers (7) are devices which can measure distance with a very high degree of accuracy using a reflected laser beam.

landslides (90) are general slippage of hillsides. These may also be triggered off by the shaking of the ground by earthquakes.

Law of Superposition (32) states that in a sedimentary sequence of rocks, the youngest rocks will be on top and the oldest on the bottom.

Law of Universal Gravitation (116) states any mass attracts every other mass in the universe using a force that is directly proportional to the product of their masses and inversely proportional to the square of the distance between them.

limnic eruption (79) also called a lake overturn, is a rare volcanic disaster in which dissolved carbon dioxide (CO_2) suddenly erupts from deep volcanic lake waters, forming a gas cloud that can suffocate wildlife, livestock and humans.

liquefaction (89) is the sudden loss in strength of soil due to application of shock such as an earthquake, causing it to behave like a liquid.

lithification (34) is the process which forms sedimentary rocks from loose material by settling, compaction and cementation.

lithosphere (119) is the rigid outer part of the earth, consisting of the crust and upper mantle.

logarithmic scale (106) uses powers of ten and the values in between such that 1.0 = log 0, 10 = log 1; 100 = log 2 and so on.

longitudinal waves (96) are those which travel in the direction of the same plane as the motion which caused it - a form of compression/tensional motion (e.g. imagine pushing over a line of soldiers who fall in the direction of the push but then recover and stand up again).

low velocity zone (120) is the place within the earth where the velocities of earthquake waves suddenly slow down and it forms the base of the aesthenosphere.

magma chamber (39) is the reservoir of molten material (magma) below the surface of the earth which may feed volcanoes.

magnitude M (105) is a measurement of the strength of an earthquake as the energy released at the source of the earthquake. Magnitude is determined from measurements on seismographs.

mantle (119) is the layer of dense rocky material below the crust.

mantle plumes (43) are very concentrated vertical currents of heat which flow from deep within the mantle to near the surface.

marker bed (21) is a layer of distinctive rock which can be seen on either side of a fault line and be used to measure the heave and throw of the fault.

Mass-wasting (27) is a general term for the movement of any regolith, Earth, material downhill.

megacalderas (64) or supervolcanoes are those with a potentially violent eruption, ejecting material greater than 10^{15} kg. They occur when magma is unable to quickly erupt through crustal rock.

meteorites (122) are stony or nickel-iron objects from outer space which strike the earth's surface.

Modified Mercalli Scale (107) measures the earthquake's intensity as a twelve-point scale of the description of the damage caused by the earth's shaking at the epicentre.

Mohorovicic Discontinuity (119) is the boundary between the earth's crust and the mantle, lying at a depth of about 10-12 km under the ocean bed and 40-50 km under the continents.

mole tracks (90) are localised ground disturbances consisting of small, interrupted uplifts and small to large escarpments due to compressional forces.

Moment Magnitude Scale MMS (107) is a measure of the strength of earthquakes introduced by Caltech in 1979 to overcome problems which the Richter Scale. It is based on the seismic moment, or the energy released when rocks move during the earthquake, which is equal to the rigidity of the earth multiplied by the average amount of slip on the fault and the size of the area that slipped.

monocline (33) is an anticline or upward fold with one limb horizontal giving only one inclined side.

nappé (32) is a complex overfolding similar to cloth being pushed over itself.

nuée ardente (39) from the French for glowing cloud, is another name for a pyroclastic flow.

orogeny (2) from the greek *oros* for mountain and *genesis* for creation, is the building of mountain ranges by uplift, usually with much folding and faulting.

oblate spheroid (127) is an elongated sphere, flattened at the tops and bottoms and extended slightly at the sides.

overfold (33) is a fold where one of the limbs has an angle greater than 90^0.

parasitic cones (48) are smaller volcanic cones formed on the sides of volcanoes by secondary eruptions.

periodotite (119) is a dense, coarse-grained ultramafic (iron-magnesium rich) igneous rock consisting mostly of the minerals olivine, pyroxene and little silica.

phreatic (55) is a violent volcanic eruption due to the presence of large amounts of steam

plate tectonics (103) is a very well-defined theory suggesting that the surface of the earth consists of many plates which are in constant motion.

plateau basalts (42) are extensive flows of basic igneous rocks which cover a very large area forming extensive plateaux. often their source are large volcanic fissures so they are often called fissure basalts.

plunging fold (31) is a fold which has also had its fold axis tilted in a direction usually at right- angles to the force which caused the fold.

ptygmatic folds (36) are small to large scale folds, usually shown in quartz or other veins within metamorphic rock showing the deformation which changed the rock.

pyroclastic flows (39) are sudden large bursts of very hot volcanic ash and gas which are blasted out of an explosive volcano, often sideways which flow down the sides of the volcano.

Rayleigh waves (96) named after Lord Rayleigh (English: 1842-1919), are earthquake waves which travel on the earth's surface and are both transverse and longitudinal waves.

recumbent fold (33) has been folded over itself and often lies at a very shallow angle.

regolith (27) is a general term for rock, soil and other earth debris.

rheidity (29) is measured as the time taken for deformation to exceed 1000 elastic limit deformations. If forces are applied for times greater than the rheidity of the particular rock, it will be deformed like a fluid (i.e. flow).

Richter Scale (106) is a logarithmic scale measuring earthquake magnitude by the amplitudes of their seismograms on calibrated seismographs.

rock falls (28) is the rapid movement of rock and soil by a sudden drop, usually due to undercutting from below.

rock slide (28) is the movement of regolith by a slipping of rock or soil, especially in layers downhill.

sand geysers (89) are produced by forces within the ground compressing the soil and causing it to erupt through the surface as a semi-liquid geyser.

scoria (49) larger, blocky material ejected from volcanoes. They often have many gas holes or vesicles.

seiches (89) are standing waves (to-and-fro waves action) caused by earthquake waves travelling through smaller bodies of water such as lakes and pools.

seismogram (93) or seismogramme, is the readout, either on a paper chart or electronic display of the various earthquake waves received by seismographs.

seismographs (92) or seismometer, are the instruments which measure vibrations in the earth such as earthquakes (and explosions made by humankind) by the relative motion of the earth's surface and a freely- suspended mass.

shadow zone (121) is an area on the surface of the earth on both sides of the planet opposite the focus of an earthquake where no body waves (p-waves and s-waves) appear because of the reflection and refraction effect due to the boundary at the Earth's core. They lie between 103° and 143° of epicentral angle from the epicentre of the earthquake (i.e. 103°and 143° between the earth's centre and the epicentre). L-waves are received at seismic stations in these zones because they travel around the earth's surface.

shield volcano (42) very large eruptive structures of rounded profile built up by successive layers of lava. They are commonly formed within a tectonic plate by a hot mantle plume below.

SIAL (120) for silica and aluminium, are rocks of the upper part of the earth's crust of lighter density.

SIMA (120) for silica and magnesium, are rocks of high density in the lower crust of the earth.

slickensides (19) are the scratches made on the face of a fault scarp by the movement of rock upon rock.

slope gradient (27) is the steepness of a surface and can be measured as an angle to the horizontal or the ration of the vertical distance over the horizontal distance between reference points.

slumping (28) is the movement downhill of regolith by a rotational motion producing curved partial layers of debris.

slump folds (34) are formed in the early stages of sedimentation when the beds are still pliable and are able to be folded under weight of the beds above or some sideways pressure.

Soil creep (27) is the ripple effect on a steep slope due to the slow downward movement of the soil surface.

solfataras (60) fumaroles which emit large amounts of sulfur vapour. In many places of the world, these are used as harvesting points for this element.

strain (10) is the amount of deformation a rock undergoes as the ratio of the increase in length, surface area or volume, to that of the original dimensions of the rock body.

stratovolcano (51) also called a composite volcano and is formed by successive layers of ash and lava. Typically, it has the well-known cone shape with even slopes. Commonly formed at subduction zones where one plate is pushed below another, re-melting crust and surface sediments.

stress (9) is the applied force per unit area on a rock and can be measured in newtons per square metres,

subduction zone (103) in the theory of plate tectonics, this is a place where one plate is pushed below another.

supervolcanoes (32) are volcanoes or large calderas which have been the site of massive volcanic eruptions ejecting more than mass greater than 10^{15} kilograms of material.

syncline (33) is a downward fold or section of a fold.

tectonic plates (83) are the mosaic pieces which form the surface of the earth and are in constant interaction and movement

tephra (39) are all volcanic pyroclastic particles ranging from extremely fine ash to very large blocks.

thixotropic (89) is a term applied to material such as some clays and clay-cemented rocks which become mobile or liquid when shaken by earthquakes.

throw (21) is the vertical displacement of the fault as seen by a marker bed.

tiltmeter (76) is a device for measuring the change in slope of a volcano's sides.

tors (5) are rounded, often balanced boulders which are formed by the chemical weathering on the surface of an uplifted intrusive igneous rock.

transverse waves (96) are those which travel in a direction which is at ninety degrees from the motion which causes them (e.g. flicking a skipping rope up and down will produce a transverse wave along it).

triangulation (100) is the process of determining the location of a point by forming circles from known points and noting where they intersect.

traveltime curves (99) are a graphs of arrival times, commonly P or S waves, recorded at different points as a function of distance from the seismic source.

tsunami (88) from the Japanese for big wave, are large waves with long wavelengths and so have massive amounts of water behind the wave front. They are caused by undersea earthquakes having a large vertical displacement.

turbidity current (29) is a sudden underwater flow of rock material down submarine canyons off the Continental Shelf and down the Continental Slope.

Universal Gravitation (Law) (125) states that the attraction between any two masses is proportional to their masses and inversely proportional to the square of the distance between them.

Volcanic Explosivity Index VEI (54) this is an open ended scale of the explosive nature of a volcano, devised in 1982 by members of the United States geological survey (USGS) and the University of Hawai'i. Its scale ranges from a value of 0 for non-explosive eruptions to that of the largest volcanoes in history given a magnitude 8.

Notes

This book is also available in electronic format which can be purchased at amazonz.com for Kindle or other electronic devices such as PCs and iPad using the free Kindle App. Books in the **ADVENTURES IN EARTH SCIENCE** series are available from Felix Publishing, Australia (info@felixpublishing.com) and include:

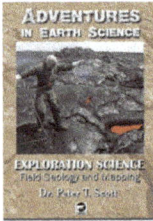
EXPLORATION SCIENCE
Field Geology & Mapping

FOSSILS - LIFE in the ROCKS

RICHES from the EARTH
Minerals & Energy

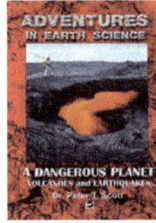
A DANGEROUS PLANET
Volcanoes & Earthquakes

CHANGING the SURFACE
Erosion & Landscapes

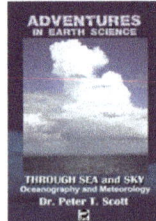
THROUGH SEA and SKY
Oceanography & Meteorology

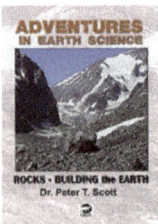
ROCKS - BUILDING the EARTH

BEYOND PLANET EARTH
An Introduction to Astronomy

www.ingramcontent.com/pod-product-compliance
Lightning Source LLC
Chambersburg PA
CBHW050727030426
42336CB00012B/1444